THE
IMPOSSIBLE
MAN

THE IMPOSSIBLE MAN

ROGER PENROSE
AND THE COST OF GENIUS

PATCHEN BARSS

BASIC BOOKS
New York

Basic Books

Hachette Book Group

1290 Avenue of the Americas, New York, NY 10104

www.basicbooks.com

Printed in the United States of America

First Edition: November 2024

Published by Basic Books, an imprint of Hachette Book Group, Inc. The Basic Books name and logo is a registered trademark of the Hachette Book Group.

The Hachette Speakers Bureau provides a wide range of authors for speaking events. To find out more, go to hachettespeakersbureau.com or email HachetteSpeakers@hbgusa.com.

Basic books may be purchased in bulk for business, educational, or promotional use. For more information, please contact your local bookseller or the Hachette Book Group Special Markets Department at special.markets@hbgusa.com.

The publisher is not responsible for websites (or their content) that are not owned by the publisher.

Print book interior design by Bart Dawson.

All images courtesy of Roger Penrose unless otherwise indicated.

Library of Congress Cataloging-in-Publication Data

Names: Barss, Patchen, author.

Title: The impossible man : Roger Penrose and the cost of genius / Patchen Barss.

Description: First edition. | New York : Basic Books, 2024. | Includes bibliographical references and index. |

Identifiers: LCCN 2024000777 | ISBN 9781541603660 (hardcover) | ISBN9781541603677 (ebook)

Subjects: LCSH: Penrose, Roger. | Mathematicians—England—Biography.

Classification: LCC QA29.P437 B37 2024 | DDC 510.92 [B]—dc23/ eng/20240202

LC record available at https://lccn.loc.gov/2024000777

ISBNs: 9781541603660 (hardcover), 9781541603677 (ebook)

LSC-C

Printing 1, 2024

For Andrea, Abe, and Isaac

CONTENTS

Contents

PROLOGUE

Courtesy of Owen Egan.

On Tuesday, December 8, 2020, Roger Penrose awoke alone in his Oxford flat. He walked barefoot and silent to the kitchen and pulled a pot of coffee from the refrigerator. He brewed coffee every Friday, stored it cold, and microwaved it cup by cup throughout the week.

As he splashed milk in his mug, he idly considered the death of the universe. The cold globules of lactose, protein, and fat spread through the hotter liquid, creating graceful clouds and tendrils. One stir of the spoon, and the shapes disappeared, leaving a mugful of monochromatic, pattern-less brown liquid.

This was the fundamental path of all things—islands of hot and cold flowing into one another, mixing and churning to create beautiful, ephemeral patterns before disappearing into a shapeless, uniform distribution of matter and energy. As time grinds up the future to make the past, the universe moves relentlessly toward greater entropy.

It will one day reach a "heat death," a state of inert equilibrium, with temperature and density the same and unchanging in every direction at every scale.

Roger had recently separated from Vanessa Thomas, his wife of more than thirty years. In his late eighties, living on his own for the first time in decades, he kept life as efficient and simple as possible. He ate plain meals of fish and vegetables or bread and cheese, with J. S. Bach or BBC Radio 1 for company.

In the months since the United Kingdom went into pandemic lockdown, Vanessa had been delivering groceries, medication, and other necessities. She texted when she'd left bags outside his door, and he pulled them in.

Every two weeks, he chopped and mixed walnuts, almonds, crystallized ginger, and shredded wheat into a large batch of muesli. He used one of his flat's two bathtubs for banana storage. He submerged them in a plastic container in the tub, keeping them underwater using a heavy orange-and-blue glass plaque he had received in 2013 from the University of Texas, San Antonio—a thank you for delivering a talk titled "Seeing Signals from Before the Big Bang" for the school's Distinguished Visiting Lecture Series.

Roger had devoted his life to understanding the past, present, and future of the universe. He was acutely aware of how every cup of coffee will eventually cool to room temperature, every chunk of stone and ice flying through space will erode into its constituent parts, every star will eventually sputter out of existence, every galaxy will collapse, and even the last surviving black holes will ultimately boil away into an oblivion of scattered radiation.

Roger didn't accept this depressing fate. The way he told the story, the universe had been around much longer than most people thought, and its distant future portended renewal and rebirth rather than a slow, gradual death. Few other physicists shared his view: Roger's theories of cyclical cosmology had left him as isolated in the world of physics as he was in his own flat.

He checked the clock on his phone. Time was progressing at its usual steady rate of sixty minutes per hour, twenty-four hours per day.

Roger, though, also knew the passage of time wasn't the brutal, indifferent entropy factory it seemed. In fact, when he found doing so useful, he questioned whether time passed at all.

People typically perceive the flow of time like the stream of words in a book—a sequential, page-by-page journey from the beginning of the story to the end. But from another perspective, all those pages exist simultaneously: readers can hold the entire book in their hands at once, even if they can only experience one word at a time. Relativity—Albert Einstein's astoundingly powerful theory that formed the basis for much of Roger's life work—treats the universe as a static object in which the past, present, and future are like the pages of a cosmic book. Time *seems* to flow from one event to the next, but viewed in four dimensions rather than three, established history and the seemingly unwritten future coexist with equal reality.

Roger found this idea of a *block universe* fascinating and convenient. He could hold all of space-time in his mind at once: the distant past and future; the inaccessible regions beyond the visible universe; the interiors of black holes and the auras of distant dying galaxies. In a block universe, the passage of time was an illusion. Aging, death, and loss shed their significance. In a block universe, Roger could pretend that his own time and energy had no limit.

He resisted any thought of slowing down—he still published and lectured like a new scholar out to prove himself. He shrugged away his fame, his best-selling books, his long, distinguished academic career. His work remained unsettlingly unfinished.

He wilfully ignored medical issues—high blood pressure, macular degeneration, mobility problems, subtle but perceptible cognitive decline—that created challenges for his work and personal life.

He could concentrate when uninterrupted, but small disruptions could throw him off for hours as he struggled to recover his train of thought. He had increasing difficulty recalling names and words and often grew frustrated with his inconsistent memory.

He had become clumsy at hunting and pecking on his laptop, which made writing painstakingly slow. He kept a piece of card jammed under the caps lock key to stop himself from accidentally depressing it and wasting hours writing paragraphs in useless all caps. Editors willingly moved deadlines for him, but no amount of time ever seemed sufficient.

His eyes had also betrayed him, blurring shapes and obscuring words on the page. He'd had a customized magnifier embedded in the right lens of his thick glasses, and by closing one eye and holding a book or paper up to his nose, he could still read. For email and digital publications, he bought the largest computer screen he could find and blew the text up to marquee-sized letters. He rolled his office chair back from the display and held a monocular provided by the National Health Service up to his eye, swivelling back and forth to read each line.

Since he was fourteen, he had kept a journal in which he jotted down ideas and drawings. He had filled dozens of books over his lifetime and still kept one within arm's reach as he researched. The once confident and graceful lines of his diagrams and sketches, though, had become shakier and less precise. His drawings looked more childlike now than they had when he was six years old.

When he lectured, he memorized the order of his slides, because he could not see them during the presentation. If a slide happened to be missing or upside down, he sometimes wouldn't know until someone told him. He had unwillingly begun using PowerPoint for some talks but still preferred the hundreds of colourful, hand-drawn acetate transparencies he kept carefully ordered in shelved stacks in his home office.

Roger ignored time as much as he could, carrying on as though he were still the same productive, creative, and inspired mathematician he had been for more than sixty years.

On a typical day, he left his dirty dishes in the sink, turned off the radio, and settled into his office to read recent scientific papers and pore over maps of the cosmic microwave background (CMB) radiation. He usually made just one outing a day: an hour-long walk along nearby canals or to the bakery in the city centre that made his favourite chocolate biscuits and multigrain bread. As a COVID-19 precaution, he held his breath each time he passed another human being.

This December day, though, was not typical. Instead of his usual rumpled shirt, oversized sweater, and well-worn, wide-wale corduroys, he chose a crisp white shirt, maroon tie, and fitted grey suit. He tamed his wispy grey hair and left his cosmology equations.

Dishes in sink, stick in hand, he descended two flights of stairs and exited to the drive where a car idled, waiting to take him into London. Reaching the Swedish ambassador's residence took a little over an hour. There, in a sparse outdoor ceremony attended by Vanessa and their twenty-year-old son Max, he accepted the 2020 Nobel Prize in Physics.

Later that afternoon he returned home and put the Nobel medal away in the same cupboard where he kept his Order of Merit, his Eddington, Dirac, and Knight Bachelor medals, and several dozen other honorary degrees, plaques, certificates, and honours.

With that out of the way, he changed into comfortable clothes and went back to work.

He won the Nobel for a 1965 paper proving that large dying stars would inevitably collapse to a point of infinite density known as a *singularity*. His singularity theorem had disrupted the world of theoretical physics, demonstrating the incompleteness of Einstein's relativity. The quest to discover new physics to describe what happens inside a black hole continues at elite research institutes around the world to this day.

Nearly fifty-six years passed between Roger's publishing his singularity theorem and the day the Swedish ambassador presented him with a 175-gram disk of eighteen-carat gold bearing the embossed image of Alfred Nobel.

His journey to the most prestigious prize in physics, though, began even earlier.

Roughly 13.7 billion years before Roger won the Nobel Prize, a hot, dense universe wrenched into existence. The new universe was tiny: all of the reality we now know as the visible universe resided in a point possibly a billionth the size of a proton. This point contained the raw material that would become light and electricity, gravity and magnetism, and the strong and weak nuclear forces. It would become matter, antimatter, dark matter, and dark energy. It would become every element of the periodic table and every particle in the Standard Model.

The chain of events that followed that moment of creation not only led to the existence of Roger Penrose but also formed all the cosmic objects and forces that guided his life's work.

Some physicists think the newborn universe contained just one fundamental force. But at 10^{-43} seconds, gravity split away, manifesting a distinct set of properties and effects. Gravity—the very same force that kept Roger's bathtub bananas under water—became the fundamental key to the creation of space, time, and life as we know it.

The next fraction of a second saw the emergence of the strong nuclear force, which binds the hearts of atoms. One theory suggests the strong force also triggered a faster-than-light expansion and cooling of the universe known as *cosmic inflation*. In a hundredth of a millionth of a trillionth of a trillionth of a second, the universe's diameter doubled more than ninety times—ending up over 1,000,000,000,000, 000,000,000,000,000 times its original size. Inflation ended as suddenly as it began in a little-understood event known as the *graceful exit*.

The still-expanding universe became a baryon-photon plasma—a primordial fog of charged particles so hot and dense that atoms couldn't form and light couldn't move. No gaps. No exceptions. Just unrelenting opacity, with not even a way to view it from "outside." There *was* no outside. The fog filled the universe, and the universe was the fog.

The fundamental forces fragmented further, leading to the emergence of the weak nuclear force and electromagnetism. Mysteriously, most of the antimatter disappeared, leaving behind a universe dominated by matter. The cosmos was well on its way to becoming the strange and beautiful expanse we observe today, filled with the quasars, quarks, photons, and forces that would become Roger's lifetime obsessions.

After 379,000 years, the universe cooled enough for the first simple atoms to form. This event, the *great recombination*, created almost all the hydrogen that exists today. It also made the universe abruptly transparent. When the fog lifted, light filled every point in the cosmos with an energy so hot and bright that it still reverberates today as cosmic microwave background radiation. Roger lived in hope that this CMB, the oldest directly observable record of the early universe, would yield up evidence to vindicate his maverick cosmological theories.

In this new light, the cosmos remained almost perfectly uniform, the same in every direction and at every scale. Without something to break that monotony, the universe might have stopped evolving, remaining as uniform and unchanging as a perfectly mixed cup of coffee.

Instead, unimaginably small random energy bursts called *quantum fluctuations* created tiny disturbances. A bump in energy here, a drop there. Gravity subtly amplified these perturbations, pulling some atoms closer together, leaving faint halos of emptier space behind. Symmetries broke. Pattern and structure emerged.

The expanding universe became lumpy.

Two hundred million years passed: these were the "Dark Ages," when cosmic radiation was everywhere, but no new light sources had yet formed.

Hydrogen clouds billions of kilometres in diameter spun as they collapsed in on themselves, compacting into whirling bulbous disks. In and in, faster and faster, hotter and hotter.

Gravitational pressure became so great, it overpowered the electromagnetic forces that had kept individual atoms separate from one another. In a burst of nuclear energy, they fused into a heavier element: helium. Nuclear fusion released massive amounts of heat, light, and other radiation. The clouds turned into fireballs. The universe blazed with the first starlight.

Gravity was at work everywhere—collapsing nebulae and forging stellar nurseries. Some of its effects were more mysterious: an enigmatic source of gravitation created larger structures—star clusters and galaxies spinning in unfathomably huge discs and spirals. This *dark matter* did not interact with light or electromagnetism, working invisibly to sculpt the largest known structures in the universe.

After billions of years, the primordial stars started to run out of hydrogen. When a star runs out of nuclear fuel, it first collapses and then explodes back out into the surrounding space, scattering atoms and plasma into a massive cloud. Gravity goes to work, slowly and steadily pulling that matter back together.

Nebulae collapsed into stars, and stars exploded into nebulae, in a continual cycle of death and birth. Larger stars crushed hydrogen and helium into heavier elements: carbon, oxygen, nitrogen, and phosphorus. Iron and calcium, zinc and potassium, silver and gold. These elements, the leftover "stardust" from ancient explosions, are the building blocks of life, the stuff of bones and bananas and Nobel Prizes.

With each nova and supernova, some material never made it back into the centre. Instead, it spun away, coalescing into planets, moons, and asteroids. Trillions of solar systems sprang up, containing swirling gas giants, rocky planetoids, and lifeless lumps of ice and dust.

When very large stars died, gravity took over completely. The stellar remains collapsed in on themselves, crushed beyond the point of no return. Matter and energy disappeared beyond an *event horizon* where time and space were so warped that nothing could escape. Each event horizon contained a singularity (as Roger would one day prove). The horizon and the singularity together became known as a *black hole*.

About five billion years ago, our own sun flared to life, surrounded by a maelstrom of proto-planets and asteroids. An ancient collision between a molten, Mars-sized planet and a rogue planetoid sent a plume of liquid rock out into space, where it congealed into a lifeless sphere with about one-quarter the diameter of the new planet it now orbited: Earth and the moon.

At Earth's surface, water existed in solid, liquid, and gaseous forms. The planet had a thin skiff of atmosphere. With these and other favourable conditions, life emerged. Atoms combined into molecules, and molecules into more complex structures like bacteria, sponges, plants, and—relatively recently—human beings.

August 8, 1931, dawned grey and humid in Colchester, England. The universe was still cooling and expanding. The sun, Earth, and moon were orbiting one another as they had for billions of years.

The baby Margaret Penrose had carried in a dark sea of amniotic fluid for nine months was nearly ready to enter Earth's atmosphere. Margaret called the family doctor to meet her at the hospital. But overcast Saturdays in August were not ideal for getting medical assistance: her doctor had gone fishing.

Margaret had studied medicine and had helped many women through pregnancy and labour. This was her second child, and with a team of maids and other servants standing by, she prepared to have the baby at home.

Around midday, her second son took his first gasps of air, registered light and shadow, felt his mother's skin.

Primordial hydrogen was in the water molecules that formed the humidity of that day. It was in Margaret's sweat as she laboured and in Roger's first tears.

Like every human being, his tiny body was made from stardust and the ashes of the Big Bang. He was brand new to the world and also billions of years old, depending only on how closely you looked.

At the atomic level, a sprouting acorn and an ancient oak barely differ: both are made from materials nearly as old as time itself. Only at more "coarse-grain" levels—those of cells, tissues, organs, organisms—do living things become young or old, inexperienced or wise, ripe or rotten. The chemistry and biology of life, death, aging, and decay mask the enduring physics beneath.

Like every newborn, Roger Penrose had the history of the universe written in each of his two trillion atoms. And like every baby, he was a living, breathing, thinking entity for whom every new movement was an experiment and every new sensation a revelation. He would grow up caught between these two realms, constantly trying to lose himself in the eternal mysteries of the cosmos but always bound to the emotional and physical vagaries of human existence.

He could never entirely leave this world behind, no matter how illusory physics told him it was. He still suffered broken bones and unrequited love, made friendships that spanned decades, experienced the pettiness of academic rivalries, married twice, sued one of the world's largest producers of toilet paper, dined with the queen of England, and rushed home on Sunday evenings to catch new episodes of *Monty Python's Flying Circus*. Much as he fought against it, he remained rooted in a reality where time passed, bills had to be paid, and decisions had consequences.

He developed a rare capacity to see beyond normal human senses, to think in four dimensions, to peer inside black holes, to bend geometry to his will, to wrestle infinity under control. But his genius didn't emerge out of thin air. His life and work only existed within the context of the people and the world around him.

For the people close to him, his desire to escape into a less ephemeral world of eternal numbers and shapes was a source of both fascination and pain. He could be self-centred and remote, justifying his distance by dismissing personal, everyday concerns as inconsequential byproducts of the limitations of human perception. He carried on emotional affairs, grew apart from his children, rejected the support and care of people who loved him, and rationalized this behaviour as resulting not from his own choices and decisions but from the blind laws of physics.

He became one of the most influential mathematicians and physicists of the twentieth century, inhabiting a "world-behind-the-world" where all the jagged fragments of human existence—heartbreak and regret, aging and mortality, estrangement and loss—disappeared, leaving behind only the clean lines and graceful curves of an endless, perfect geometrical universe.

He devoted his life to questions that straddled physics, art, and philosophy: Why is the universe like it is? Why is there something rather than nothing? Does beauty point the way toward truth?

What is the nature of consciousness? Is mathematics something we invent or discover? Is there something deeper and more fundamental than space and time? To this last question, Roger's answer was an emphatic yes. Space and time—and every other aspect of the physical universe—emerged from something even more essential. At the deepest level, all the subatomic particles and galaxy clusters, energy and mass, heat and the light disappear, revealing the primary source of reality: the universe's transcendently beautiful mathematical skeleton.

The universe, Roger believed, was made of geometry.

Elite physicists say Roger's work was like "magic"—that his ideas seemed to emerge more from epiphanies and miracles than from the normal drudgery of incremental scientific advance. Non-scientists, who revere his optical illusions, impossible objects, and other hypnotic geometric oddities, treat Roger with a cultish awe. He had an uncanny ability to find beauty and meaning in the shapes of things—be they puzzle pieces or the invisible arcs of four-dimensional space-time. His life's work was more than merely aesthetically pleasing: again and again, he revealed the "unexpected simplicity" of the machinery of the universe, inspiring wonder among scientists and non-scientists alike.

His intensely visual sensibilities made him a renowned and beloved public speaker, a hero of puzzlers and recreational mathematicians, and a respected thinker, even when he relentlessly worked to disrupt the agenda of theoretical physics and other sciences.

Roger paid a price for his success, and so did many people around him. Not only did he alienate large swaths of the global scientific community, but he also grew distant from family and friends. He placed his work ahead of all other concerns and left a trail of heartbreak and anger in his wake.

In his late eighties, disconnected from friends, family, and colleagues, he lived mostly in silence and solitude. He continued to be bathed in honours and accolades, to wander through the universe looking for new truths and deeper answers, and to race against time to understand the story of reality.

1

THE JUNGLE

On a midday in mid-July 1937, six-year-old Roger Penrose perched on his seat at the end of the heavy wood table that dominated the Penroses' large dining room. The table could easily seat ten: in the evenings, visiting scientists, artists, and family friends filled the spacious, high-ceilinged room with noise, smoke, and conversation. At lunchtime, the room was airy and quiet. Tall-back chairs and wide doorways made the house feel like a giant's castle.

Legs dangling, crossed arms like sticks, Roger looked hardly older than his four-year-old brother, Jonathan, and much younger than eight-year-old Oliver.

He regarded a scoop of stewed greens prepared by the family cook. The green blob was the only thing standing between him and the freedom to disappear into the hot, sunny afternoon.

Lionel and Margaret—the Penrose boys only ever called their parents by their first names—didn't eat lunch with their children. Instead, Isabella Black, a twenty-year-old Scot doing double duty as maid and nanny, oversaw Roger's painfully slow meal.

Sunlight filtered through the high paned windows, projecting slow-moving parallelograms onto the clean wooden floors. The windows looked out toward Lexden Road, a wide, straight street running through an affluent neighbourhood in southwest Colchester. The two-storey house sat well back from the road. Trees and shrubberies muffled the traffic. The back garden was wild, shady, and lush. It sloped downward to a high hedge with a gate onto a secluded, grassy footpath running parallel to the street behind the row of houses.

His brothers had long ago finished their food and melted away into the summer heat, leaving Roger alone with Stella (as the boys called her) and the greens.

Neither was getting any friendlier.

"If you eat half of it, that's alright. Then you can go," Stella said.

Roger glowered at the green circle. The list of foods he enjoyed eating—topped by marmite spread on chocolate biscuits—did not include boiled spinach.

Stella tidied, waiting for Roger to budge. With her attention momentarily elsewhere, he at last picked up his spoon, reshaping the spinach into a tidy semicircle. Not a leaf left the plate.

"Well, I've done it," he said.

She looked at the plate. He looked at her. A tiny flicker of mischief passed over his face. Whether charmed or just worn down, Stella conceded.

"Alright then," she said. "You may go play."

This was Roger's first lesson on the power of geometry.

Roger clambered down from his chair.

In a collared shirt and striped braces, short pants hiked up over his navel, his thin frame seemed small and vulnerable. But his slightly crooked, curious smile betrayed a sense more of adventure than danger.

During the week, his father either worked at his office at the Royal Eastern Counties Institution for the Retarded or sequestered in his home office poring over papers and tables related to his research into the genetic origins of "mental defects."[1] He was doing research for the Colchester Survey, a multi-year, career-defining project on what would later come to be known as Down syndrome.[2]

Margaret was equally unavailable, busy with household matters and other unspecified commitments. Servants came and went, cooking, cleaning, and tending, but Roger had the gardens and the world beyond all to himself.

Roger often followed the path behind the house eastward and then north, gravity leading him further down the slope into a small, grassy glen. Where the land started climbing again, the path entered a dense wood of oaks, field maples, and hornbeams.

Roger and his brothers called these woods "the Jungle." Forking dirt paths pierced the trees and underbrush, crossing small, shady clearings and bursting out at the top of the hill into the sunlit meadows of Hilly Fields Park, with views of Colchester Castle and the town centre. A child could easily get lost in these woods. Roger made a point of doing so frequently. At any moment, he imagined, a wild tiger or elephant might emerge from behind a tree or a spaceship swoop down through the canopy. The wood covered only a few acres, but Roger disappeared into the Jungle for hours at a time, lost amidst the towering trees and his own tall tales.

One bright, still summer day not long before he turned seven, he emerged from the leafy shadows of the Jungle into a bright, quiet clearing. Sunlight warmed an old stone bench. Nearby a pillar rose out of the grass. It flared out at the top, but the structure was too high for Roger to see its top.

His feet left the ground. He rose gently through the air until the top of the pillar was at eye level. He was not as alone as he had thought: his father was there, unexpectedly hoisting Roger up to see the plate fastened to the column. Lines and crosses formed a ring of symbols along its perimeter. A compass rose marked out the cardinal directions. Along the north-south axis, a thin metal wedge angled up out of the centre, as though an arrowhead shot out of the sky had lodged in the plate. The metal cast a sharp shadow on the markings along the edge of the disc.

Floating in Lionel's arms, Roger sensed something unusual in his father's demeanour: happiness. Lionel generally behaved indifferently toward his sons. He had an aversion to shows of emotion and a near-total lack of interest in the lives of his children. But on this unusual occasion, holding his son up to the face of a sundial, his normally serious expression melted into a smile. Something extraordinary was happening.

Lionel's emotional vacancy was more than generic British stoicism. Displays of emotion had disgusted and alienated him since he was a child himself.

Born in 1898, Lionel was the second of four boys. His mother, Elizabeth Josephine Peckover, came from a family of wealthy Quaker bankers and minor nobility. She married James Doyle Penrose, an Irish portrait artist. They bought a large estate known as Oxhey Grange, near Watford, a small town about thirty kilometres from London. Oxhey's Victorian mansion anchored more than 400 acres of farms and fields, which became the childhood backdrop for Lionel, Roland, Alexander, and Bernard.

Elizabeth and James taught their sons traditional Quaker values: love of God, speaking truth to power, appreciation of fine art, music, and scientific knowledge, a deep commitment to pacifism, and—especially—avoidance of strong shows of emotion. Lionel absorbed it all, except for the belief in God. His brothers grew up to become hedonistic artists and world explorers, but Lionel dedicated his life to rigorous pacifism and science.

Slim and just under five and a half feet tall, Lionel hardly bore an imposing physical presence. But he observed the world with intense, intimidating curiosity. He studied rather than enjoyed it.

During World War I, Elizabeth and James allowed young Quaker men to work their fields and sleep in their outbuildings rather than join the conflict.[3] Lionel, though, chose a much more punishing path of pacifism.

In late 1917, he trained as a volunteer for the Friends Ambulance Unit, created by British Quakers a year earlier to provide humanitarian relief to British and Allied soldiers. The unit moved wounded and ill combatants by train from the front lines to hospitals across France.

By June 1918 he was travelling back and forth across central and northern France with the No. 5 Ambulance Train. Each day, he wrote one or two lines in a diary, noting logistical information and glossing over the horrific details of his work.

July 11: Load at Gézaincourt for Rouen: A night-trip passing through Abbeville. Lurid sunset.

July 13: A.M. Clean windows etc., at Abbeville. P.M. Raid. Several bombs on Abbeville—one near the train, which smashed windows previously cleaned. Went to dugout.

July 14: Wait all day—walk out onto hills in evening to avoid raid and get rain instead.

August 7: Dysentery cases... Start loading 1 a.m. Finish 4 a.m.

August 9: Load 1 a.m.–7 a.m. French wounded. Very bad cases. Wounds not dressed.

Near midnight on October 31, his ambulance train paused outside Givenchy-en-Gohelle for emergency repairs. A coal train crashed into the back end of the No. 5, sending stretchers, patients, and volunteers scattering off the rails.

"No one hurt. Patients took it all fairly well," he concluded.

Travelling through Boulogne, Saint-Pol-sur-Mer, and Rouen, the unit kept moving patients while making repairs on the fly. On November 9, they took the ambulance train into Paris for final repairs. He was still in the city on November 11, 1918, the day the war ended. On Armistice Day, he finally released his feelings of horror and disgust.

He took aim, though, not at the ravages of war but at the raucous celebration of peace.

> At 11 a.m. cannons fire and church bells ring madly. The armistice has been signed. I had come to Paris with several others, but broke away. The streets very rowdy, everybody chucks work. Kids let out of school go parading the streets, headed by their masters. Girls sell flags of all sizes to passersby. People stand at their windows and shout and wave flags. Flags of the allies magically sprout from every window, chiefly French and American. Old ladies—French officers—Yanks—girls—old workers—children form small processions in the streets to rush up to one another and kiss or shake hands.
>
> My curiosity goads me to wander for miles through dense intoxicated crowds until I too am utterly sick with the smell of spirits, the crushed flags, the broken glasses, the glistening eyes of the crowd, the hideous laughter, shouts of men and women turned into lunatics—this is peace.

After the war, he studied mathematics at Cambridge. He graduated in 1921 and spent a year abroad at Vienna University, studying psychology and attending lectures by Sigmund Freud. He underwent psychoanalysis himself but quit when he found it "led to the acquisition of a quiet effrontery."[4]

Adept at chess, piano, woodcraft, sketching, and other arts, he was a polymath adrift until a chance meeting halfway up a mountain in the Austrian Alpine district of Zell am See put him on a new path.

There he met twenty-five-year-old Margaret Leathes, a fellow Cambridge student whose own meandering path had brought her to that same windy Alpine slope.

Unlike his father, Roger's mother wasn't actively taught to bury her emotions. But life offered her plenty of reasons to do so.

Margaret's father, John Beresford Leathes, a physiologist and pioneering biochemist, was a second son in a long line of clergymen. His parents expected him to study for the priesthood, but he discovered atheism at an early age and, to his parents' great disapproval, pursued a scientific career instead. Margaret's mother, a Jewish concert pianist named Sonia Marie Natanson, had fled Russia in the late 1800s "with a hatbox full of roubles and little else."[5] John and Sonia met in Switzerland where Sonia, a natural polyglot, worked as a language tutor. She helped John with the German he needed to understand the lectures he was attending. They married in Switzerland and returned to England.

John and Sonia's first child, a boy, died at birth, strangled by his own umbilical cord. When Margaret was born, her parents nicknamed her "Bob." They never forgave her for being a girl.

Sonia maintained no ties with her relatives in Russia, and John's family cut him off when he left the church. When Margaret was young, John had only a £150-per-year student stipend; Sonia earned a little money translating the books of Ivan Pavlov, the Nobel Prize–winning Russian physiologist. Margaret primarily remembered Pavlov's visits for his "beard like Father Christmas."

By eight years old, Margaret had discovered a love of performance. She played Mark Anthony in a school production of *Julius Caesar* with such passion that she brought herself, as well as the audience, to tears.

The family moved to Canada in 1909, where John lectured at the University of Toronto. Sonia was active in the Canadian suffrage movement. Outspoken on women's rights, she kept her Jewish heritage hidden, including from her daughter.[6]

Margaret frequently found herself in the company of globally renowned scientists, activists, and artists. She recalled composer and pianist Sergei Rachmaninoff visiting them in Toronto, idly playing the same upright Broadwood piano she used for her daily practice. "I remember his haircut which looked as though his scalp had been mown with a lawnmower," she wrote in an unpublished memoir. "But I don't remember how he played."

Sonia was also friends with suffragette Emmeline Pankhurst and her daughters Christabel and Sylvia. "Inspired by them, I used to deliver impassioned speeches about votes for women standing on the washstand of the spare bedroom," Margaret wrote. "Sylvia once spent the night at our house in Toronto about 1910 as a guest of my mother. She gave me a goodnight kiss which I shall always remember."

Sonia and Margaret returned to England in 1913, a year before John joined them. They found a flat in Chelsea, and Sonia enrolled Margaret at Bedales Independent School, founded in 1893 as "a humane alternative to the authoritarian regimes typical of late-Victorian public schools."[7]

Humane may have been an overstatement.

Even in winter, the students bathed daily in unheated water in their unheated dorms. The only reprieve came when the air turned so cold the water in the metal tubs froze solid, making bathing impossible. The girls went on a forced four-mile run each morning at 6:30 through the dark and cold. "Despite the fact that all this Spartanism was supposed to increase the circulation, most of us developed very bad chilblains," Margaret wrote.

She thrived nonetheless, excelling at competitive chess, piano, acting, and swimming and diving. She aced exams and charmed her teachers, eventually ascending to the status of head girl.

Bedales was coeducational, but girls and boys were kept apart, and expressions of sexuality were forbidden. "Feeling unplatonic was quite against the school rules. Such feelings had to be deeply hidden, if not fiercely smothered for the benefit of the 'Chief' (our name for the headmaster)," she wrote. "Though, of course, nearly everyone really had such feelings."

Margaret spent a year at Sheffield University, where her father taught, studying history, physics, and philosophy. In 1920, she moved to Cambridge to study medicine. She was one of two women studying alongside 200 "rowdy ex-servicemen, fresh from the 1914–1918 war. They made too much noise for concentration. They stomped when one came into the room—in time with one's footsteps—and when one sneezed or coughed they cheered," she wrote. "I once left the lecture room in despair, but...was reproved for having done so."

She learned to take up less space, make less noise, and elicit as little notice as possible.

She reconnected with a friend from Bedales, Eileen Rutherford, daughter of famed nuclear physicist Ernest Rutherford. "I frequently spent time at her home in Queens Road and listened to Sir Ernest and Lady Rutherford quarrelling, while I played on the pianola," she wrote. "I even went for walks with Sir Ernest and was told that the important thing about being a scientist was to have 'good hunches' otherwise one would become buried in irrelevant detail."

She lived at a hostel in Gordon Square, where she befriended the writer Frances Marshall, who took her to parties that included "all the famous Bloomsbury characters," including Vanessa Bell, Virginia Woolf, and Ralph Partridge. In such august, creative company, Margaret could not escape feeling like "a mere fly on the wall." She never believed that any of these marvellous writers and artists found her interesting or important enough to remember.

To celebrate and mourn the end of her Cambridge years, Margaret took a hiking and sightseeing tour through Austria. Two weeks into the trip, she gathered with other Brits at the base of Grossglockner, the highest mountain in Austria.

There she met a small, serious man, a friend of a friend, named Lionel Penrose. "No bells rang, no sparks flew," Margaret recalled.

Margaret and Lionel were roped with other partners. A biting wind picked up as they climbed, whipping icy rain against the climbers. Wet, cold, and miserable, they turned back before even attempting the summit.

Two days later, Margaret boarded a train back to England. To complete her medical training, she had arranged to take on an internship at the Royal Free Hospital in London. Lionel, still half-heartedly studying psychology, happened to be on the same train. Noticing an anatomy textbook in Margaret's lap, he peppered her with questions and, by the time they reached England, had decided to pursue a medical degree himself.

He showed up at her door a few days later with an orchid sticking out of a chianti bottle. They ate dinner together that night—and almost every night thereafter.

Lionel completed three years of medical school in one, fully qualifying as a doctor just three months after Margaret.

They married in October 1928. Margaret continued to take short-term positions at clinics around London, building her experience and network. The days were long and tiring, but she rose to each challenge. In just the first week after she qualified, three friends each independently sought from her an urgent abortion. On a house call during a stint at the Bermondsey Medical Mission, she correctly diagnosed a twelve-year-old boy with an appendix abscess just in time to save his life. She wore her copy of the hospital pharmacopeia to shreds treating women and children all over London.

Lionel, who had a public reputation as a supporter of women pursuing careers in science and medicine, found it increasingly unbearable to watch Margaret succeed. She took a position at an ophthalmology clinic in New Cross that required her to work evenings. He insisted on accompanying her. Patients complained that Margaret looked too young to be trusted as a medical professional. These complaints, plus Lionel's overbearing presence, became too much. Margaret quit the clinic. She never practiced medicine again.

Within five years, she had three sons, a house tended by a team of servants, and no employment. Her days of acting, performing music, and speechifying were long behind her—Lionel denied her every professional, emotional, and creative outlet. Meanwhile, he ascended into the medical career she had once planned for herself.

His reputation grew steadily, leading in 1931 to a high-profile appointment as research director at the Royal Eastern Counties Institution, which brought him and Margaret to Colchester.

He began the Colchester Survey the year Roger was born. Over the next seven years, he evaluated 1,280 patients with mental disorders, their 6,629 siblings, plus parents and other relatives.

He enlisted Margaret to help him collect information on patients and their families. Together, they created massive data tables on parental age, birth order, stillbirths, incidence of abnormalities, and instances of similar conditions manifesting among siblings or more distant relatives, which brought rigour and evidence to what had traditionally been a highly subjective branch of medicine.

Despite how fiercely he valued his own intelligence and that of his friends and family, he did not accept a prevailing, brutal belief that "mentally deficient" people were also morally deficient. He refuted arguments for sterilizing medically diagnosed "idiots," "imbeciles," and "morons," both by illuminating the moral bankruptcy of such eugenic ideas and by demonstrating their impracticality. He unflinchingly trusted his own knowledge and insights, even when they placed him outside mainstream scientific ideas—a trait his middle son would also come to exhibit later in life.

Through the entire project Margaret was at his side, collecting, sorting, and analysing data.

Lionel publicly acknowledged and praised Margaret's contributions in his publications. But that enthusiasm disappeared whenever she dared try to do anything on her own.

With no paid work and with hired staff to care for the house and the children, Margaret had more than sufficient free time to return to competitive chess. When Roger was seven years old, she registered for a championship tournament in London—her first competition in more than a decade.

The night before the tournament, Roger awoke to hear the unusual sound of his parents' raised voices. He lay awake late into the night listening to them argue, as Lionel railed at Margaret for having entered the competition.

The next morning Margaret was too exhausted to leave the house. Lionel had worn her down. She did not compete.

This was how Roger understood his parents: Lionel was serious, curious, and humourless. He was always in control, especially of his wife. Margaret, after a life of repression, was very apt to relinquish her autonomy and restrain her emotions. "My mother was very non-demonstrative in her love for us. She would do anything for us. But there wasn't this sort of warm hugging love about her," Roger recalled.

Part of her was simply missing—fiercely smothered for the benefit of the Chief.

Roger was born into a family of privilege and wealth. Once it found a foothold, though, his sense of emotional isolation grew more entrenched with each new experience and memory.

His first-storey bedroom overlooked the front garden. Late one summer night, yowling cats just outside the open window shattered the peace. Their cries first entered Roger's dreams and then disconcertingly carried on when he awoke. Terrified, with no idea what was happening, he screamed for help.

"Lionel! Margaret!"

When they didn't come, he called for his older brother, Oliver.

"I screamed and I screamed and I screamed and nobody heard me. Nobody came," he recalled more than eighty years later.[8]

A blinding light shone through his window, moving across his bedroom wall as if there had been a prison break. Police sirens joined the cacophony.

Neighbours had heard Roger's screams, thought someone was being tormented, and called the police, who arrived on the scene before his own family. "My parents and the cook and the maids and Oliver didn't hear me. When they did come in eventually, it was because the caterwauling didn't stop, and because of the police coming by and shining lights in through the window."

Roger believed he was on his own.

Lionel and Margaret took Oliver, Roger, and Jonathan on a road-trip holiday. Lionel had arranged for them to stay in the former home of British poet Rupert Brooke in the village of Grantchester, a couple of hours' drive from Colchester. He and Margaret drove separately, dividing their children between two cars.

Halfway through the return trip, they stopped at a restaurant. A quick inventory of their sons revealed that Roger was nowhere to be found—each parent had thought he was in the other vehicle. They rushed back to Grantchester to find him sitting in the garden drawing circles in the dirt, unaware he had been abandoned.

With such low expectations of his father, Roger found their shared discovery of the sundial strange and exhilarating.

He forgot about elephants and tigers. He placed his hand on the warm plate. The metal wedge was sturdy and unyielding, but Roger could not grasp the sharp line that divided light from shade. The shadow seemed to jump through his hand when he tried to grab it. No matter how quickly he covered one hand with the other, the shadow always ended up on top. The device's only moving part seemed to obey its own magical rules.

Lionel helped Roger decode the Roman numerals and other markings. He told Roger how the ancient Greeks had given the central wedge its name: *gnomon*, "the one that knows." Shaped and positioned correctly, the shard of metal could indicate the passage of time nearly as effectively as a clock's gears, springs, and pendulum. The gnomon pointed due north, its upper edge tilting away from the dial at an angle of fifty-two degrees—Colchester's latitude—demarcating a line parallel to the axis of the Earth's rotation. The planet itself was the cog whose turning allowed the shadow to mark time's progress.

Roger watched the shadows move, imagined Earth turning beneath them, considered the light source blazing millions of kilometres away. His house, the gardens, and even the Jungle and its imaginary wildlife shrank to become tiny specks, as Roger's mind opened to the larger workings of the solar system's cosmic machinery.

He didn't yet have the words (or the mathematics) to describe a two-dimensional shadow marching across a three-dimensional structure marking movement through four-dimensional space-time, but he was absorbing his first hints about the ways shape, shade, light, motion, and time were connected. His father's uncharacteristic joy heightened the impact of the discovery. He was glimpsing something so amazing, even Lionel was moved by it.

The world turned. Shadows grew longer. The clockwork universe indicated it was time to go home.

In the months that followed, Lionel and Roger returned many times to the sundial. To Roger's happiness and great surprise, his father's excitement did not abate. Lionel's work grew no less demanding, his emotional vacancy no less pronounced, but still he found time to teach Roger how to read the dial with more sophistication, showed him how the shadows changed with the seasons, and explained the geometry and mechanics of the sundial in greater detail. Roger was awestruck. His connection to Lionel and to the wider world grew together. Roger took from it a powerful lesson in how to feel not quite so alone in the universe.

2

UNEXPECTED SIMPLICITY

Clouds of cigarette smoke and beery arguments drifted around the dining room, half-obscuring the crowd of scientists, artists, mathematicians, and obsessive oddballs who'd found their way through Lionel and Margaret's front door. The Penroses welcomed any and all fascinating people at their table. They provided wine and food, and guests brought new scientific ideas and theories, impromptu musical performances, puzzles, and challenges.

Some guests were acclaimed researchers. Others just needed hospitality. Some stayed for a single meal. Others didn't leave for months.

Lionel's generosity was almost compulsive. He shared his home, food, and money without hesitation or concern for the cost. Intellectually, he was fierce and stubborn. In social settings, he couldn't say no. Every meal brought together a new combination of old friends and passers-through, girding themselves to match wits with whoever else happened in.

The jousting was convivial, but everything happened under Lionel's scrutiny—he owned the table and the conversation.

Roger habitually installed himself in a quiet corner of the dining room, listening late into the night. That nebula of smoke, music, and debate held the essential building blocks of his future intellectual life. Much of what he heard flew over his head. But he absorbed fragments and proto-ideas that would lie dormant for years before bursting into flashes of brilliance later in his life.

Lionel and Margaret burned brightest in Roger's mind: Lionel's unrelenting curiosity and obsessive generosity, his fierce scepticism, his emotional vacancy, his easy shifts between academic research and mathematical play; Margaret's withdrawal and deference to Lionel, her love of grammar, her steadfast support for her sons' interests and pastimes, her tendency to observe rather than participate. Part cautionary tale and part role models, Roger's parents prized intellect and mental talent above all else. "The three boys were terribly emotionally deprived. All they got rewarded for was exceptional performance," his first wife, Joan Penrose, later recalled.[1]

Roger didn't feel deprived. With a sharp mind and a watchful eye, he soaked up everything he could from his parents and their friends. Hugh Alexander, a chess master and a cryptanalyst, American mathematician Norbert Wiener, philosopher Freddy Ayer, and many others dined at 47 Lexden Road. Roger stored their conversations like unused puzzle pieces, ready to pull out later as needed.

Famed British mathematician, code breaker, and computing pioneer Max Newman had been Lionel's roommate in Vienna in the 1920s. Lionel and Max bonded over their shared love of the piano, chess, making and breaking codes, brainteasers, and puzzles. Lionel was best known for chess puzzles, while Max was more famous for abstract logic.

Max's puzzles were merciless and dazzling, even for elite solvers. His best-known logic puzzle, Caliban's Will, is almost hateful in its stinginess of information:

When Caliban's will was opened, it was found to contain the following clause:

I leave ten of my books to each of Low, Y. Y., and Critic, who are to choose in a certain order:

1. No person who has seen me in a green tie is to choose before Low.
2. If Y. Y. was not in Oxford in 1920, the first chooser never lent me an umbrella.
3. If Y. Y. or Critic has second choice, Critic comes before the one who first fell in love.

Unfortunately, Low, Y. Y., and Critic could not remember any of the relevant facts; but the family lawyer pointed out that, assuming the problem to be properly constructed (i.e., assuming it to contain no statement superfluous to its solution) the order could be inferred.

What was the prescribed order of choosing?[2]

Solving Caliban's Will required a meta-understanding of logic puzzles—one had to understand certain things to be true even though they couldn't actually be proved within the puzzle itself. Roger would one day write books about the non-computability of human understanding based on this exact type of insight.

At age eight, the solution was beyond him. But he still marvelled at how difficult Max's puzzles were and how clever the creator had to be. As with the sundial, these challenges sparked palpable joy and amusement in the adults around Roger. Roger soaked up these moments of delight, feeling them deep in his tiny bones.

Lionel's puzzles matched Max's in brilliance and brutality. He was known for "two-mover" chess puzzles—quick challenges with just enough twists to confound all but the most expert solvers. He also created mathematical puzzles, 3-D jigsaws, and visual brainteasers to entertain and test those in his orbit.

Roger sensed the edge underlying his father's gregariousness. Puzzles weren't merely intriguing; they were for sorting those who could pass a challenge from those not up to snuff. Intellect, the surest path to winning Lionel's approval, became fundamental to Roger's sense of self-worth.

Margaret did not join the competition or debates—it was not worth sparking Lionel's anger by flexing her own intellect. She kept the peace by keeping her thoughts to herself.

When Lionel and his friends were not around, Margaret came alive. She taught Roger to love language and to delight in the precision of grammar, spelling, and usage. Like Lionel, her enthusiasm came with judgement: poor English indicated poor character.

The branches of the Penrose family tree were laden with artists, scientists, and other accomplished thinkers. Roger needed to find a way to take his place among them.

Experiences like the sundial and the spinach resonated with something powerful inside him. Of all his intellectual and artistic pursuits, geometry awakened the strongest feelings. Shapes were intuitive, powerful, and beautiful. The phases of the moon, the slope of the garden, the configuration of steel strings inside the family piano: every shape brimmed with hidden meaning, telling a story about how things work and why things are the way they are.

It wasn't enough to see shapes and patterns—Roger also was driven to create them. The precise, clean lines of his childhood drawings already exhibited hints of the sensibilities that would define his later scientific diagrams.

As a gift for his portraitist grandfather James Doyle Peckover, Roger glued into a banner six sheets of paper torn from one of his father's old notebooks. He covered them with an elaborate pencil-drawn road system. The banner slid through two slits Roger cut in a seventh sheet, also covered in roads, creating an evolving map of kaleidoscopically disconnecting and reconnecting lines. "This is a sort of cinema. When you move the sliding thing along to each line, the picture changes," he explained in a note to his grandfather.

Roger tapped into the geometric realm that would provide him—and his friends, family, and fans—with a lifetime of delight and awe.

Mathematical ideas stuck in Roger's mind. Less so, politics and current events. He was barely aware of the growing threat of war in Europe and his parents' increasing fear.

In 1938, the Friends Ambulance Unit was preparing to resume operations, but Lionel was now in his forties with a young family, an inherited fortune, and a thriving career. He had no interest in witnessing the death and devastation of war so closely again.

Max Newman's talent with logic and puzzles would take him to Bletchley Park, where he led a team of code breakers dedicated to decrypting Axis messages. He tried to convince Lionel to join him, but Lionel wanted no part in the war. He wanted only to get his family as far from the conflict as possible.

About fifteen kilometres north of Colchester, there was a centuries-old estate comprising about 700 acres of farmland anchored by a rambling, derelict home called Thorington Hall. The original eight-bedroom structure was built in 1630 and had passed through many hands over the centuries. By the time Lionel set eyes on it, the manor, baking house, stables, and other outbuildings were long abandoned and barely standing.

Lionel had bought Thorington in 1937 and poured money into its restoration. Workers removed a lean-to that farm labourers had built along the north side. They unbricked fireplaces and repaired the massive six-flue chimney that nearly doubled the height of the three-storey building. They preserved original architectural details, including elaborate hand-carved oak newel posts and ceiling timbers chiselled to look like multifaceted jewels. A sliding door hid a secret "priest hole" once used by Catholic clergymen to escape persecution.[3]

Lionel had the brick walls around the garden repaired and a tennis court installed behind the stables. A small stream flowed below the east-facing windows, and in every direction tramping paths led through hilly farmland and woodland. Even with rattling, draughty windows, sagging doors, and broken masonry, Thorington became a tranquil haven in a country tumbling toward war.

Lionel, though, deemed it insufficiently remote to serve as a refuge for his family.

He donated Thorington Hall to the National Trust, an architectural preservation charity. The Penroses retained rights to use the hall, though it would be years before they did so. Instead, Lionel offered it as a wartime refuge for London seniors who lost their homes in the Blitz. By the time the first displaced residents arrived, the Penroses were 6,000 kilometres away.

Lionel secured invitations to lecture on genetics and mental health in Columbus, Ohio, and Philadelphia, Pennsylvania. On April 14, 1939, the Penroses departed from Southampton on the *Aquitania*, a Cunard White Star ship bound for New York.[4] The ship manifest listed Lionel as "Doctor of Medicine," the three boys as "scholars," and Margaret as "housewife." Her academic and medical credentials were long forgotten.

Staying at the Greenhill Farms Hotel near Philadelphia, Margaret enrolled Roger, Oliver, and Jonathan in Friends' Central, a Quaker school next door to the hotel. They stayed for several months until Lionel found a position as a hospital physician in London, Ontario. The family waited out the war in this small Canadian city.

On September 3, 1939—the day Britain and France declared war on Germany—the family settled into a comfortable rented house on a wide boulevard near the Thames River and the University of Western Ontario, where Lionel took a position as a lecturer.[5]

Many other wealthy Brits also spent the war years in North America. Roger, Oliver, and Jonathan were three of more than 5,000 child "war guests" who ended up in Canada.[6]

Lionel and Margaret received many requests from family friends in England, asking for their own children to join the Penroses at their Canadian haven. They made arrangements to rent a hotel for the summer, planning to house and care for nineteen British boys and girls.

But when the German navy began sinking passenger ships crossing the Atlantic, many parents changed their minds. Only two children ended up living with them.

Margaret's childhood friend, a former ballet dancer named Penelope Barman, put her son Benjamin on a boat to North America. Margaret met Benny at the London train station. He stood alone on the platform with a Paddington Bear–style name tag hanging around his neck. "He was happy to see me and was not, apparently, unduly disturbed," Margaret recalled. "He...exhibited but one feature of concealed homesickness which was that he had a pathologically enormous appetite. It had to be seen to be believed."[7] Lionel and Margaret welcomed him as family, and Roger gladly shared a room with him.

A few months later, another family friend, Edith Hogben, descended on the Penroses' home with her two children, Clare and Julian. Edith and Julian moved on to Ottawa, but Clare stayed with the Penroses for the next two years.

Life in Canada brought mountains of snow in winter and sweltering weather in summer. With no servants, Margaret learned to feed the household herself.

Roger turned mealtime into another exercise in time dilation and geometry. He made Margaret cut four triangles off a slice of corned beef, turning a square into an octagon. When he had eaten the triangles, she pared the remainder down to a diamond, feeding Roger the scraps. Then a triangle. Smaller triangles. Finally lunch would be over. Roger's love of shapes saved him from starvation.

The mealtime quiet of Lexden Road was also long gone, replaced by a chaotic mix of Benny's appetite, Oliver's obsession with algebra and calculus, Jonathan's fanaticism for chess, and Clare's accounts of school life. It all swirled around Roger as he pretended to eat. Margaret wryly captured the scene:

Teatime—1000 Wellington Street, 1941

MARGARET: Now hurry up Roger, Benny has nearly finished his third helping and you haven't begun yet.

ROGER: Oh, but I haven't said grace yet: I can't in this noise.

OLIVER: Margaret, shall I tell you some more about integral calculus? I have just worked out a new equation: dy by dx...

CLARE: You know us girls in our class were passing round a *hoojumaflip* paper with personal remarks about everyone on it. I nearly died when I saw mine. It said, "She's a swell kid but boy, what ghastly nail polish."

JONATHAN: I said knight to queen bishop four, what would black do then?

LIONEL: Bishop to king's three.

JONATHAN: You can't, there's a pawn in the way.

LIONEL: True, I wondered whether you'd remember it. Queen takes pawn.

BENNY: I say, Lionel, I've got a *good* idea.

ROGER: Do seventeens go into 571? I can't decide whether it is a prime number or not.

MARGARET: Jonathan, don't pick your nose. Benny, don't use your glass as though it were a telescope. Roger, you've only had one mouthful, take another.

ROGER: There's bits in it.

BENNY: I say Lionel, I've got a *good* idea. Why don't we send over a big aeroplane and bump off Hitler with a big bomb, boom bang, bang!!

LIONEL: Oh shut up, everyone.

OLIVER: There seems to be a snag in my equation, shall I tell you what it is? You see...

BENNY: Can I have another piece of currant bread?

MARGARET: I set the limit at seven slices.

BENNY: Well, can I have some more cake then?

ROGER: Benny, I don't believe you ever said grace.

CLARE: Of course there's a rule that boys and girls can't walk down the darned corridor together, but I've walked down with dozens of boys.

OLIVER: What, all at once?

CLARE: No one at a time silly, and no one has ever said anything—maybe that's 'cos I'm English.

ROGER: But if you took a dodecahedron with the corners partly cut off and you...

MARGARET: Roger take another mouthful for God's sake.

ROGER: There's bits in it.

BENNY: Please can I have some more honey, this pot is empty.

JONATHAN: What would Black do then?

MARGARET: He wouldn't pick his nose I'm sure.

BENNY: I say Lionel. I've got a *good* idea.

JONATHAN: Margaret, will you buy us Jumbo comics?

BENNY: And buy me Fantastic comics?

ROGER: And buy me Weird comics? But I must pay for it out of my own pocket money, I won't take it if you give it to me.

OLIVER: And buy me *Analytical Geometry*.

LIONEL: Shut up, everyone.

Margaret kept her journal hidden from Lionel.

In Canada, she managed to cultivate a social life and even revived some of her old pastimes in ways that didn't seem to spark Lionel's ire. She kept a chessboard set up by the telephone. Local calls were free in Canada, and she spent many hours playing chess over the phone with a friend across town. Roger would listen to them play as he carried on with his own drawings and crafts.

Roger and Benny's bedroom was on the second floor, overlooking the lawn and neighbouring houses. Each morning, they strapped roller skates onto their shoes, burst outside, and careened downhill to the public school a few blocks away. Roger invented a system he called the "Victoria Street Road Clock," which entailed counting the students he passed each day in order to calculate how late he was for class. If the streets were very empty, the hour was cataclysmically late.

In class, Roger initially struggled, especially in math. He had no trouble understanding the concepts, but it took him too long to get through exercises. He doodled instead of calculating, and his wandering mind continually got in the way of completing assignments. He often ran out of time, handing in half empty tests and frustrating his grade-three teacher until she dropped him back a year.

Oliver had already skipped a grade, which made Roger's setback smart all the more. Grade-two math proved much too easy for him. A few weeks later, he arrived at school to find himself mysteriously moved up into an advanced grade-three group with a different teacher. He assumed that Miss Watson had simply wished herself rid of him.

Things changed in grade four. "I think you could do better in these tests if you had more time. I'm going to let you have as much time as you like," Mr Stinnet told him.

Roger soon found himself once again sitting alone, long after the other children had vanished. He relished calculating at his own speed, as unrushed and steady as a gnomon's shadow. Time had been the only thing holding him back. He could finally think beyond individual answers and feel out the ways numbers, angles, and equations really fit together. Mr Stinnet taught him how the digits of any multiple of nine always added up to nine. Roger learned the "alternating sum" test to determine whether a number was divisible by eleven. He took these tricks apart, tinkering with them until he understood how they worked as thoroughly as a master mechanic understands a car engine.

He moved to the top of the class. But the rewards that came with unlimited time went much deeper.

Though he wouldn't formally name it for another thirty years, Roger had discovered one of the most powerful motivating factors of his career: "unexpected simplicity."[8] Human existence was complicated, messy, and confusing. It was filled with cranky teachers, emotionally distant parents, war, and awkwardness. But mathematical statements had such breathtaking clarity, Roger could only marvel at how their intricacies melted away, revealing the raw beauty beneath.

He didn't care that two plus two equals four. But he was hypnotized by the fact that two plus two equals two *times* two. This odd equality between the sum and the product rattled around in his head for weeks until he discovered a second pair of numbers that fit the same pattern: $3 + 1\frac{1}{2} = 3 \times 1\frac{1}{2}$.

He had to tell Oliver. "He was a little older than I was and considerably more able at mathematics. After a little calculation, he produced the fact that any two numbers whose reciprocals sum to unity have the property that their sum is equal to their product," Roger recalled decades later. "He then showed me how to arrive at the result algebraically. I was fascinated by this. It was the first time I had ever seen how results could be simplified by means of algebra. It was extremely mysterious and beautiful to me."[9]

Roger had found two pairs of numbers that fit a pattern. Oliver had generated an infinity of them using one simple formula. "Any two numbers whose reciprocals sum to unity . . ." For Roger, the language of mathematics read like poetry and felt like magic.

Where else was unexpected simplicity hiding? Did mathematical beauty relate to physical reality? Did people invent these rules and formulas, or did they discover them already existing out there in the universe?

On another day, Roger studied the backsplash of the kitchen sink, tiled in pale blue hexagons, tessellating with perfect regularity. He imagined extending the pattern all along the wall and even further.

"Lionel," he asked, "could you cover the whole world in hexagons?"

"You can't do it with hexagons, but you can with pentagons," Lionel said.

Soon the two of them were sitting on the floor cutting shapes out of cardboard and exploring how some—triangles, hexagons, squares, rectangles, parallelograms—could cover a flat plane in regular, repeating patterns known as *tessellations*, while other shapes, like pentagons and dodecagons, left gaps and irregular holes that kept them from fitting together.

Lionel showed Roger how the geometry changed when they tried to tile curved shapes, like a sphere or a torus. A typical football, for instance, comprises twelve pentagons and twenty hexagons stitched together to approximate a sphere.

Together they cut and folded twelve identical cardboard pentagons into a dodecahedron. They turned piles of triangles into tetrahedrons, octahedrons, and icosahedrons and made six squares into a cube. With the dodecahedron, they now had a complete set of Platonic solids—three-dimensional shapes with identical sides, faces, and vertices.

Lionel taught Roger the Greek prefixes and suffixes he needed to name each polygon and polyhedron. After the Platonics, father and son worked through the thirteen Archimedean solids, assorted rhombic polyhedra, and increasingly complex cardboard constructs as elaborate and delicate as cut gems.

When they tired of three dimensions, Lionel introduced Roger to polychora, four-dimensional analogues of polyhedra. It was physically impossible to build such shapes, but Roger had no trouble imagining a fourth axis jutting at right angles to height, width, and depth. Working now in his mind rather than with his fingers, he combined cubes, pyramids, and spheres to construct four-dimensional objects. Doors opened in his mind. The hexagonal pattern he'd begun with was part of something larger and more magnificent than he had ever imagined. And as he explored that elaborate multidimensional geometry, his father was once again with him, guiding him, sharing his excitement.

Oliver often read to Roger. At the height of his polyhedra years, they discovered Robert Heinlein's sci-fi short story "And He Built a Crooked House." In it, a Los Angeles architect constructs a house with eight cubical rooms arranged in a four-storey, three-dimensional cross. An earthquake causes the house to slip and transform into a tesseract—the four-dimensional analogue of a cube. Roger could imagine the architecture so vividly, it was as though he was actually walking through it. Roger knew the physical world had no fourth spatial dimension, but the mathematical concept was so powerfully intuitive, he couldn't help feeling as though it had its own reality.

Roger and his brothers had many bonds and common interests, which often manifested as a rather cutthroat sibling rivalry.[10] In their competition, they couldn't even leave games of chance to chance. Jonathan consistently, mysteriously outplayed Roger at Rock Paper Scissors. Hour after hour Jonathan just seemed to know what Roger would throw before he knew himself. Did he have a tell? Did he follow some pattern only Jonathan could see?

Emulating his father, Roger took extensive notes, documenting his and Jonathan's moves each round, searching for some telltale predictability in his own choices. When he found nothing, he created and memorized a table of randomized throws. By sticking to the table, he pushed Jonathan down to an acceptable 50 percent win rate.

Something else struck Roger: How did Jonathan react so quickly to make winning throw after winning throw? How did his brain work so fast? Was his brother conscious of what he was doing, or was it somehow happening below the surface?

Roger, Oliver, and Jonathan strung a badminton net between two trees beside the house. Roger dug a precise rectangular trench in the lawn, delineating court boundaries. Every lightning-quick drive, flick, and drop-shot mystified Roger. Brains seemed to work more quickly than physics should allow.

Table tennis was even faster. As he did his best to smash and back-spin his brothers into submission, Roger was observing the absurd speed at which they all acted and reacted. He stowed this mystery of the high-speed brain away with other questions queued in the back of his mind.

His brothers were excellent subjects for the study of brain activity. In chess, both boys could checkmate Roger before he knew what was happening. When playing Concentration, Jonathan instantaneously memorized the location of every upturned card, while Roger struggled to find a single pair.

When Jonathan started outperforming him on violin, Roger switched to viola to avoid competition. With Oliver playing flute and Lionel piano, they formed a family quartet. Lionel recorded their performances in wax to share with friends. Margaret was not permitted to join them.

When Roger didn't feel like competing, he disappeared into solitary artistic fantasies. He created elaborate pop-up books depicting a daring space adventurer named Q13 Raygun. Q13 blasted across the pages, keeping Earth safe from the Octopus from Mars and other interplanetary villains.

The family's patch of Canadian lawn frequently transformed into the surface of an alien planet, field of battle, or unexplored desert where Roger's boyhood heroics saved many innocent victims from imagined plane crashes, monster attacks, and enemy invasions.[11] He built models of Spitfires, Mosquitos, and other fighter planes to stage battles and dogfights that would have horrified his pacifist parents.

One hot summer day, Roger wanted to pitch a tent in the back garden and sleep outside overnight. He approached Lionel, who sat behind a large wooden desk, poring over papers. He did not look up when

Roger entered his office. Roger's voice came out sounding both thin and intrusive.

"Lionel, could I have permission to camp out in the back garden?"

Lionel looked up from his papers, devoid of interest or affection. He stared briefly and witheringly at his son and returned to his research without uttering a word. When the silence became unbearable, Roger fled.

The experience permanently scorched Roger's memory. He saw the pattern.

In the sharp shadow of the sundial, his father was animated and engaged. When they talked about tiling and polyhedra, Lionel was a loving father. But for anything else, Roger was not worth the breath it would take to respond. Hot with embarrassment, he resolved never to bother his father with such trivia again.

Lionel occasionally travelled to Ottawa for work. He left the boys with puzzles to entertain them while he was away.

"If you look at a cube from a certain angle its silhouette is a perfect hexagon," he said to them before one trip. "Your challenge is to figure out an analogue for a tesseract."

Oliver and Roger filled page after page with pencil sketches, trying to find a perfectly regular, symmetrical three-dimensional section of a four-dimensional cube. They folded cardboard into various shapes, trying to imagine what kind of shadow the tesseract might cast in our reality.

Ultimately, they failed to find a solution. When Lionel returned, he retrieved a length of stiff wire and bent it into a rhombic dodecahedron—a twelve-sided figure with diamond-shaped faces—turning it this way and that so the boys could see how it formed a symmetrical slice of the higher-dimensional solid.

These moments were magic. Roger understood everything about the shape—beautiful in its own right, part of something larger and unseen, and also somehow connected to his father.

Roger learned to use cleverness not only to win affection but also to show it. He bent the tip of a wire coat hanger into a semicircular dial and mounted it on a toilet paper roll to make a "moon clock" for

Benny. He showed Benny how to look through the tube, position the moon in the centre, and rotate the hanger dial to match the moon's phase. The angle of the tube and the position of the wire revealed the time.

Roger was smaller than Benny but more athletic than his physique let on. He seized every opportunity to fight gravity, scaling walls and climbing trees. Around the corner from his makeshift badminton court, a huge old hardwood towered over the Penrose home, its spreading branches brushing the roof.

He easily ascended until he could see into his and Benny's second-floor bedroom. Heart thumping, he edged out further, feeling the bounce of the branch with each step away from the trunk. One wrong move, he knew, and the ground would be accelerating up at him at an increase of thirty-two feet per second for every second he fell.

He made the jump, scrambling onto the sloped roof. He slipped through a dormer window into his bedroom. From then on, he came and went via the upstairs window as often as through the front door.

Lionel fully immersed his family in rustic Ontario life. Not long after they settled in Canada, he bought a decrepit cottage in the village of Bayfield, about an hour's drive from London and a few minutes' walk from the cliffside shore of Lake Huron. A sign on the sagging porch read "Blink Bonnie," a Gaelic phrase roughly translated as "glimpse of beauty."[12] If the name had ever been non-ironic, it was long before the Penroses bought the cottage. Thanks to bad plumbing and rampant mould, local kids called it "Stink Bonnie." It was as ramshackle as Thorington Hall, with none of the history or faded grandeur.

Summers in Bayfield were not luxurious. With only a hotplate and no icebox, Margaret toiled to put meals on the table. The mildew and dust never abated, the small yard was perpetually unkempt and weedy, and the mosquitos were vicious.

For Roger, it was paradise.

On weekend evenings, he walked down Chiniquy Street to Pioneer Park, where a huge, white-painted wooden cinema screen teetered at the top of a cliff overlooking Lake Huron. Every gust of wind threatened to send it tumbling into the lake. The sun sank behind the screen and shadows stretched across the field until twilight dimmed sufficiently to start the film.

He sat in the grass for cliffhanger serials like *Perils of Nyoka*, relishing lurid adventures in distant exoticized locales.[13] After a show, he walked back beneath the stars to Blink Bonnie, alone and alive, as happy as a boy could be.

Back in London, Ontario, during the school year, he and Benny often visited the local roller rink. Many of their classmates also went, including a Canadian girl named Marilyn Kendal.

Marilyn was smart and outgoing. Roger found friendship with her easy and natural. They claimed desks next to each other in grade five. She always said something marvellously clever to make Roger laugh.

"Algy met a bear," she whispered to him while their teacher's back was turned. "The bear met Algy. The bear was bulgy. The bulge was Algy."

Neither of them could stop laughing. They lost control so completely that the teacher swooped in and separated them for the rest of the year. That on its own might not have killed their friendship.

Benny Barman finished the job.

Boys and girls weren't supposed to be friends, he told Roger flatly. If Roger liked Marilyn, there was something wrong with him. This kind of peer-enforced gender division was common, but Benny's disapproval left Roger feeling raw and embarrassed. How had he missed this rule? Why were some patterns so obvious to him and others so opaque?

"Somehow there was something wrong about being friends with Marilyn Kendal," Roger later recalled. "It was made out to be wrong to have a friendship with girls my own age. I think this probably affected me. [After that] I was very timid around girls."[14]

With slightly different timing or context, Benny's words might have been quickly forgotten. But as it happened, they became foundational.

Roger started avoiding girls and focused instead on building his bond with Benny. Together, they began making up a new series of adventures for Q13 Raygun featuring a new intergalactic villain named Queen Marilyn.

In 1943, a British colleague of Lionel's, Jack Haldane, wrote to alert him that the Galton Chair in Eugenics at University College had recently been vacated. Haldane, who sat on the chair selection committee, implied the position was Lionel's if he wanted it.

"I think that you and I are the British people under 60 who have contributed most to human genetics, and therefore one of us should have the chair. As you have specialized on man and I have not, your claim is somewhat stronger. So I should be very glad to know if you would like the chair were it offered to you."[15]

Lionel began to make plans to return to England. Even though all three Penrose boys were excelling at school, their father worried they would still be behind when the family returned to England.

He hired a private Latin teacher, the wife of one of his university colleagues, to keep them current. After an hour of ablatives and genitives, she relaxed and chatted with the boys.

"What do you intend for your career to be?" she asked Roger.

"Well," he replied, "if something happens, it happens."

"Oh!" she said. "Do you believe in determinism?"

He realized that he did.

His conversations with Oliver, his growing understanding of physics and causality, and his natural instinct to try to understand the ways of the world through indifferent, immutable patterns left him feeling that life was something that happened to him rather than something he controlled or affected. He couldn't beat Jonathan at Rock Paper Scissors by making different choices, only by randomizing his throws. His could navigate his parents' personalities, but he could not affect them. His choice of friends was governed by rules as immutable as those that determined the phases of the moon. Even though his mind was constantly active, time carried him along inevitably and passively. His future, he concluded, was as determined as his past.

"Space-time is out there," he told his teacher. "And I am exploring it with my life."

In 1945, at age forty-four, Margaret gave birth to her and Lionel's fourth child. Shirley Penrose was born in London, Ontario, just a few months before the family returned to England. Thirteen-year-old Roger was delighted to have a baby in the house, and especially to have a sister. He doted on Shirley and helped care for and tend her. If he couldn't be friends with girls his own age, he could at least shower love on his baby sister.

By summer, Lionel's Galton Chair professorship was confirmed, and the war in Europe was over. Two days before Roger's fourteenth birthday, the Americans dropped Little Boy, a fifteen-kiloton bomb, on Hiroshima. Three days later a second atomic bomb destroyed Nagasaki.

Though Roger knew the name Albert Einstein and understood the basics of relativity, he did not connect Einstein's most famous equation, $E = mc^2$, to atomic weapons. His parents did, though. Lionel and Margaret were horrified by the loss of hundreds of thousands of civilian lives and disgusted with the scientists who had weaponized Einstein's physics.[16]

On September 8, six days after Japan surrendered, the Penroses went out for a last lunch with Canadian friends before getting the train to Toronto, where they would board the *Samaria* for the eleven-day return voyage. They returned to Wellington Street to retrieve their packed bags from the front hall, only to find the front door locked and their landlord standing out front. He demanded extra payment for damage to the side yard—Roger's badminton-court trenches now ran between the Penroses and their voyage.

Minutes ticked by and tempers rose. Roger sweated, knowing this was his fault. Unnoticed, he slipped away from the argument to the other side of the house. He scaled a tree, skipped to the roof, and slid open his bedroom window. A few minutes later, he opened the front door in triumph. The Penroses pushed past their landlord, grabbed their bags, and fled for England.

3

THE ARROW OF TIME

Back in England, the Penroses lodged with Lionel's brother Roland while they sorted their affairs. They didn't stay long.

Roland was England's most prominent surrealist, a renowned collector and creator of surreal art. He was a close friend of Pablo Picasso and lived a much wilder life than Lionel could stand.

Shortly after they settled in, Roger recalls the household being shaken awake at four in the morning by someone throwing beer and wine bottles at the windows. "There was a tremendous noise, and a woman screaming, 'I hate all the Penroses! I hate the Penrose family!' She was the former wife of Lionel's youngest brother, Bernard. She was completely drunk. My uncle hid in the house until the police dragged her away," he said.[1]

Lionel, disgusted with his brothers' rowdy chaos, moved his family to a house on Rodborough Road in Golders Green, a central London suburb. Roger enrolled at a grade school affiliated with University College London, where Lionel had become the Galton Professor of Eugenics.

Lionel wrote and spoke publicly against eugenics, informed by his extensive research on Down syndrome. He eventually convinced the university to change his title to "Galton Professor of Genetics," which eliminated the offending term but retained the name of the man who coined it.

Oliver began calling his father "Prof," but it was too late for Roger—he could never call his parents anything more informal than their first names.

Thorington Hall emptied out after the war. It was still close to derelict when the Penroses started enjoying its shabby tranquillity for summers and holidays. In hot weather, Roger tramped through surrounding farmland and played tennis with his brothers and family guests. In winter, chill winds seemed to blow straight through the mansion, rattling cracked windows, shaking doors and beams, and slamming branches from gnarled trees against the roof and walls. Centuries of history plus abundant quirks and strange noises had fostered many stories of ghosts haunting different rooms. Burn marks in the plaster recorded where a former owner performed fiery rituals to ward off evil spirits.

The central chimney opened into a cavernous fireplace on the east wall of the grand parlour. Lionel often sat at a writing table on one side of the fireplace, jotting research notes and ideas for papers in a notebook. Watchful Roger hauled a small chair and a lap desk inside the arch of the fireplace. Seated by the warming blaze, he emulated his father, filling his own notebook with page after page of tetrominos and nonagons, tiling puzzles and doodles, and other half-formed ideas all drawn with the graceful surety of a natural-born geometer. In the fires of Thorington, he forged his lifelong habit of populating book after book with sketches and diagrams, giving his highly visual mind a new outlet for play and exploration.

Thorington was a joy for Roger, with all the space, silence, and solitary adventure he craved. Even as a teenager, the crumbling stone-and-brick walls around the property seemed to soar above him. He used toe holes and gaps to clamber up and over. At the top of a partially collapsed wall, he rigged a rickety platform from scrap wood and old metal pipes. Here he turned a pair of binoculars skyward to watch Spitfires and Mosquitos scream their way to and from a nearby military base—thrilling real-life versions of the toys he had played with in Canada. On his unsteady perch, he felt like he was defying gravity, soaring high above the world in one of those planes.

Thorington was several kilometres from the nearest village, surrounded by woods and farmland. On moonless nights, the sky blazed with starlight, which had travelled billions of light years to create this dazzling display.

Roger stepped outside one clear night to find Lionel hunched over a midsize telescope mounted on a tripod. He invited Roger to peer through the eyepiece. He'd trained the telescope not on the stars but on a nearer neighbour: the planet Saturn—just 1.5 billion kilometres away. The sight of the gas giant, its rings, and the smaller spheres of Titan, Rhea, and Tethys left Roger speechless. "This was an amazing thing for me. I had seen many pictures of Saturn. But seeing the real thing, seeing that it really looks like that—it took my breath away," Roger recalled.[2]

Airplanes needed huge, noisy engines and high speed to just stay aloft. But this ancient space traveller, with 95 times Earth's mass and 764 times its volume, had silently, weightlessly glided through space under its own momentum for billions of years.

As he stared, Saturn and its moons crept closer and closer to the edge of the telescope's viewing area. Roger knew enough about the solar system to understand this drift was caused not by Saturn's movement but by the turning of the Earth.

Lionel momentarily took over and adjusted the telescope to keep Saturn in view. In the starlight, Roger could see his father's excitement at Saturn's mysterious beauty. Doors opened, connecting Roger once more to his father and the cosmic expanse beyond. Saturn's gravity kept its moons in orbit, while the sun did the same for Saturn, Earth, and the other planets. Gravity also made his wall-topping observation platform so thrillingly dangerous and explained why airplanes had to be so loud and fast. Gravity was slowly collapsing the west staircase at Thorington and also preventing the whole solar system from drifting out of the Milky Way. But perhaps gravity's most powerful effect was the way it brought his father alive and drew the two of them together.

In these rare moments—studying the sundial, making polyhedra, staring at Saturn—he could forget his father's withering looks, his control over Margaret, his disdain for emotion. The complicated, unpleasant aspects of his world disappeared. What was left felt like perfection. He would stop time right there if he could. But time was its own mystery that Roger wasn't yet ready to master.

Thorington Hall also provided Roger with more down-to-earth joy. He loved playing tennis with his brothers and visitors. He loved wheeling the dry line marker around the perimeter of the court, creating clean, straight chalk rectangles—as geometrically satisfying but less damaging than the trenches of his badminton court back in Canada.

He remained fascinated with the speed of brains and bodies. He was convinced he processed information and made decisions more quickly than physics should allow. Oliver or Jonathan would send a ball whizzing over the net. He assessed its velocity and spin, noted his opponent's movements, felt the breeze. He could process all the data, sort through multiple options of forehand, backhand, speed, and spin, and choose a response in time to move his body to the precise location and position to make the return. It didn't make sense that could all happen in a fraction of a second.

Roger strongly suspected the mind didn't work the way he'd been taught in science class.

He had no mental space for spirits or souls; the mystery required a physical explanation. But nothing he had learned so far satisfied his curiosity about how, exactly, they could think so quickly. He periodically turned the enigma over and stored it away for future examination.

In London, Lionel and Margaret settled back into a life of dinner parties and late-night debates about genetics, chess, and politics.

Lionel and Roger hunkered down by the radio together to listen to Cambridge astronomer Fred Hoyle give talks on BBC Radio about the steady-state universe, relativity, and other current topics in theoretical physics.

When listening to Hoyle or talking to Lionel or Oliver, Roger's senses stretched in every direction, to the planets, stars, and galaxies, back to before Earth existed, and forward to long after it had disappeared. He noted the fraction of a second between light's bouncing off the faces of his family members and hitting his own retinas and the slightly longer fraction of a second before his brain became aware of that perception. Nothing happened when or where it seemed.

The sound of Hoyle speaking, a change in his brother's facial expression, the physical presence of his father sitting by the radio—everything had to move through space-time before it could affect him. And because nothing moved faster than the speed of light, those effects could only happen so quickly. When events happened close enough in time but far away enough in space, they could not be causally related.

Roger discovered one of the great geometric tools of relativity: a light cone. Light cones were spare, simple diagrams—just a few lines on a page—that compressed four dimensions into two, to create a map of every possible past event that could have affected a moment in space-time and every possible future event that could be affected by it. Time moved along the vertical axis—from the past at the bottom of the page to the future at the top. Space extended out horizontally along a single line or over a plane.

By convention, physicists used distance and time units such that any particle travelling at the speed of light moved up the page through space and time at a 45-degree angle along the exact edge of the cone—a path known as a *light vector* or *null vector*. Anything moving slower followed a steeper path within the cone.

Because nothing travels faster than light, any event outside the cone could not have a causal relationship with that moment. Roger imagined his own past light cone stretching back through cosmic history, capturing an ever-larger swath of events that led up to his own

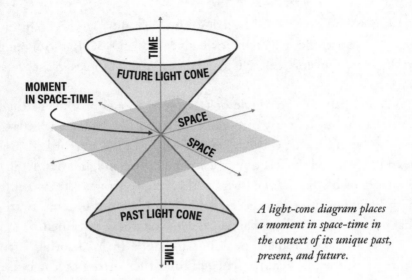

A light-cone diagram places a moment in space-time in the context of its unique past, present, and future.

existence. He pictured his future light cone, containing all parts of the universe his own actions might affect. He considered the regions of the universe outside his light cone—worlds and events he could never influence or be influenced by. All of it felt viscerally amazing. Albert Einstein's theory allowed him to know the universe in its entirety—even regions of space-time he was causally cut off from.

Roger was again held back a year at school, this time because his Latin wasn't up to British standards. And once again, he impressed his teachers sufficiently that they slingshotted him forward two years, making him one of the youngest and smallest boys in his year.

His closest school friends were three Jewish boys, who embraced him as one of their own. "I think somehow they accepted me as being a brother in some way, through my mother's mother being Jewish. The rule is the Jewishness goes through the female line. So that was supposed to be making me in their opinion Jewish."[3]

Oliver skipped grade school altogether in England, going straight on to study physics at University College when he was sixteen. Roger admired him greatly. Oliver still lived at home and continued to read to Roger from popular books about relativity, quantum mechanics, and other current physics theories.

In the 1940s, few people outside physics and mathematics departments understood relativity. The main concepts could be stated in plain language, but they defied intuition: gravity isn't a force but a product of the universe's hidden four-dimensional geometry; time slows down and distances compress for objects travelling near the speed of light relative to other observers; massive objects also slow the passage of time; empty space has a shape; energy and mass are different forms of the same phenomenon; events that are simultaneous from one perspective happen at different times from another; the passage of time itself might well be an illusion.

With relativity, rock-solid human experiences of gravity, time, movement, and space all became illusions of our limited brains and senses.

Quantum mechanics—the physics of the very small and very cold—was even more counterintuitive: particles could appear to be in multiple places or states at once and could teleport from one location to another. They could behave like waves and particles simultaneously. Quantum particles could be *entangled*, meaning their physical properties were linked, even if they were nowhere near each other. Merely observing a quantum system changed its nature.

Relativity and quantum mechanics were also mutually contradictory. A baseball cannot be in two places at once. But the quantum particles it comprises appear to have that ability. The ball obeys different rules depending how closely you look at it. This fundamental incompatibility frustrated and intrigued scientists more than any other problem in theoretical physics.

Roger and Oliver puzzled over these ideas, trying to cram their imaginations inside an atom or burst them out into four-dimensional space-time.

Together they devoured *Mr Tompkins in Wonderland* and *Mr Tompkins Explores the Atom*. In these pop science books, physicist George Gamow depicted relativity, quantum mechanics, and other scientific ideas through the dreamed adventures of the main character.

Mr Tompkins absorbed scraps of theoretical physics from lectures and chance conversations, before nodding off and journeying to dream realms where these ideas manifested as palpable, everyday experiences. In the "land of relativity," Mr Tompkins witnessed relativistic effects merely by riding a bicycle. Time sped up and slowed down; streets stretched and compressed; simultaneity came undone.

Mr Tompkins explored the subatomic world not as a quantum physicist but as a bemused traveller wandering among electrons and atomic nuclei. The books heightened Roger's sense that mathematical and physical theories existed in a world he could physically visit and explore.

He'd glimpsed elusive fragments of this realm before. Every sundial, moon clock, and telescope became a portal through which he could spy some part of that other world. Gamow's books showed him how exciting—and how much fun—actually entering it would be. "I am sure that their magic was responsible, to a very considerable extent, for the great excitement that fundamental physics has held for me for the rest of my life," Roger wrote in an introduction to a 1993 reprint of Gamow's books. "I still vividly recall the tigers of the quantum jungle, and the old woodcarver's boxes of mysterious coloured balls (the nucleons), the relativistically flattened bicycle, and the professor calling out 'Just lie down and observe' as he and Mr Tompkins see their miniature universe collapse inwards upon them. It was Mr Tompkins who made the new physics vivid and real for me."

Roger became captivated with the idea of a block universe—a static, four-dimensional, relativistic universe in which time does not flow. In a block universe, the immutable past, the ever-shifting present, and the seemingly unwritten future are all part of the same eternal existence. A block universe made free will and personal responsibility appealingly illusory: if the arrow of time was an illusion, then Roger's mistakes were not his own and merely a product of the blind laws of nature.

He found it deeply appealing to inhabit Einstein's four-dimensional space-time, like a real-life Mr Tompkins, directly experiencing relativity instead of trusting the math. This other world was beautiful and intriguing, sensible and predictable, and increasingly preferable to the messiness of teenaged boyhood.

Roger, Lionel, Oliver, and Jonathan went for long walks in the woods around Thorington. Lionel strode along alone, carrying a plastic chess set. Oliver walked ahead and Jonathan behind. Roger was the "runner," ferrying back and forth between the others, reporting his brothers' moves to Lionel, who updated the board and served as referee. Oliver and Jonathan played in their heads, mentally tracking one another's positions without seeing Lionel's board.

Jonathan had an extraordinary memory and intuition for chess. He would later become England's youngest, most successful chess player. Oliver strove to keep up by reading countless books on strategy and history. He could never match Jonathan and ultimately settled for merely becoming a Cambridge chess champion.

With Lionel's guidance, they played Kriegspiel, a punishing chess variation in which players only know the locations of their own pieces and must deduce their opponent's moves. They played Rifle Chess, Random Chess, Losing Chess, and whatever else Lionel threw at them.

Roger couldn't compete with either of them and didn't bother trying. He searched elsewhere, though, for moments that would spark his father's joy and attention. Lionel still showed no interest in making small talk or speaking about personal matters. His areas of paternal interest were narrow and consistent.

Lionel overheard Roger telling Margaret he was to begin calculus at school the following week. Lionel put down his work. "I will teach you calculus," he said.

The two of them sat down together, as Lionel animatedly walked him through derivatives and integrals, showing him how to calculate the rate of change in velocity for an accelerating vehicle and find the area between two curves. Roger sensed Lionel's pleasure in their lessons and a certain pride of ownership—he had to introduce these concepts to his son before a teacher had the chance.

• • •

Lionel and Margaret were counting on Roger to study medicine.

"I was going to be a doctor. Oliver obviously wasn't. He was going to be a physicist or something. And they gave up on Jonathan. He obviously wasn't going to be a doctor. He was just interested in chess. Chess was his life."

His mother told him he had a good bedside manner, but when Roger imagined becoming a doctor, he didn't picture sitting with ill patients.

"I really believed not that I was going to be a general practitioner—make people healthier and so on. What I wanted to do was be a brain surgeon. I want to see what was going on with this blasted thing and open it up and peer inside, and get a better idea what's going on there."

The headmaster at University College School required each sixteen-year-old to discuss with him which courses they'd take in their final two years.

"I'd like to do biology, chemistry and mathematics," Roger told him, thinking he could satisfy his parents' wishes and also keep up with the subject he loved most.

"No, I'm afraid you can't do that combination," the headmaster said flatly. "If want to do biology, you can't do mathematics. If you want to do mathematics, you can't do biology."

Roger was caught off guard. His decision in that moment would not only determine his focus for the rest of secondary school but also shape his more distant future. He wasn't used to choosing like this—he usually just let the universe unfold around him. If he dropped biology, he could not be a doctor. But he couldn't abide the idea of dropping mathematics. Was it even possible to defy his parents this way?

"Oh...Alright...How about chemistry, *physics*, and mathematics?"

"Okay, I'll put that down."[4]

The conversation lasted less than two minutes. It eradicated his medical career.

His parents were furious. That Roger and his brothers would carry on the Penroses' privileged status as a family of intellectuals and artists was assumed. They had specific expectations for Roger, though. They were not only disappointed that he wouldn't be a doctor but disgusted that he might become a physicist.

"They accused me of keeping bad company because one of my friends wanted to be a nuclear physicist. They said, 'Oh, nuclear physics is atom bombs. You're not allowed to do that. That's terrible.' It was horrible."[5]

Even for committed pacifists, the leap from enrolling in high school mathematics to facilitating nuclear annihilation was big. Roger wasn't thinking that many moves ahead. He knew he'd given up one career but hadn't thought through the alternatives. He certainly had no plans to create weapons of mass destruction. He was just trying to hold on to the mathematics that brought him such consistent joy.

"I still believed I wanted to be a doctor when I was walking up to see the headmaster, but I didn't want to lose the maths. I loved the subject. I just didn't want to lose it. It had nothing to do with my friend who wanted to be a nuclear physicist."

Two years later, Roger dropped another bombshell.

"The decision came for what I should do when I go to university. I wanted to do maths. I really loved maths and that was what I clearly was the best at. My father didn't approve at all because he thought these people who are professional mathematicians are peculiar, unworldly people. He had somebody who worked with him who was a bit like that. He was a good mathematician but a little bit strange in a worldly way," Roger said.[6]

This time Lionel worried less about Armageddon and more about Roger becoming a socially awkward oddball. He believed pure mathematics was for people who had no other skills. It indicated intellectual and social failure.

"You can do your mathematics in some other area and that's fine," he told his son. "Then you can enjoy yourself with your maths. But to actually study it as a subject on its own? No."

Lionel conscripted a highly admired math professor friend at University College named Hyman Kestelman to test Roger into submission.

"Should Roger really do mathematics?" he asked his colleague. "He really seems to be determined to do it. But is he any good?"

Lionel was looking for a reason to deny Roger's ambition, but Kestelman, who loved teaching as much as he loved mathematics itself, approached Lionel's question with genuine interest. He developed a dozen advanced problems requiring sophisticated mathematics and lateral thinking and presented them in person to Roger.

"You have a day to work on these problems," he told Roger. "You may find you can only do one or two of them. That's fine."

"The problems were a little bit strange, but interesting. They were certainly not things I'd come across in my geometry or algebra classes," Roger said.

He answered every question, and by the end of the day, he and Kestelman were chatting about Roger's theories about Fermat's little theorem and its relationship to prime numbers and Pascal's triangle. Lionel had to relent. Roger enrolled in the maths program at University College, which allowed him to continue living with his parents.

Winning a disagreement with his father felt strange and empowering. Lionel immediately began working to regain control.

"Look, you should go to Cambridge. Don't do your degree at University College," he told his son a few months into his program.

"I don't want to go to Cambridge. I'm quite happy with what I am doing in London. And suppose I do get in there—I would just be starting again."

"Well, it counts more when you're in Cambridge. Maybe you can get a scholarship to go up there."

Lionel's former family fortune had greatly diminished, but he was still sufficiently well off to pay for Roger's education. Winning a scholarship would prove Roger belonged at Cambridge. He hoped his son could at least win an "exhibition," a modest financial award with less status than a full scholarship.

Roger reluctantly allowed Lionel to drive him up to Cambridge for the exams.

"I ended up being just below exhibition level. One of the reasons was, I couldn't finish one of the papers because I made the mistake of drinking too much coffee that morning and I needed to go to the loo. It just wrecked my chances. I couldn't concentrate."

This "mistake" provided Roger with exactly what he wanted. The universe won the argument for him, proving to Lionel he didn't belong at Cambridge. He carried on living at home and studying at University College London.

Roger had trouble socializing in his undergraduate years. He was likeable—overtly clever, playful, optimistic, and talented. But he knew no more about how to talk to women than he had while drawing Q13 adventures with Benny Barman. Thoughts of intimacy, romance, and sex overwhelmed and confused him.

He had a trick for surviving social events. Working long hours in Lionel's garden shed, he had designed and built several three-dimensional jigsaw puzzles. When assembled they formed Platonic solids—a tetrahedron, octahedron, and cube. Broken apart, they became jagged jumbles of spikes and angles. One of his math professors told him only his tetrahedron puzzle was sufficiently challenging to be interesting. Roger bought a block of Perspex and, using his father's hacksaw and broken file, created a translucent version of the puzzle.

He brought his Perspex tetrahedron with him to parties and other events so he always had something in his pocket to share and talk about.

Loneliness and desperation might have overcome his awkwardness had he not found it so satisfying to escape into mathematics. Mathematics become for him a "sex substitute," connected closely to his ideas about masculinity and intimacy.[7]

When I was in my teens until mid-twenties, I had very little in the way of relationships with girls. This was to some extent just a matter of shyness on my part, but not entirely. I have always had difficulty imposing my own will in a situation or even just making a decision....I would be afraid of making suggestions often, since these might not be acceptable to the person I was with....

I somehow felt that although I was missing something by having so little in the way of such relationships, I was gaining something else. By pouring my life into my mathematics (primarily) I was able to find release for those psychic energies which would otherwise have found outlet in more normal pursuits.

And without such a period in my life I should have been that much the poorer when it came to developing those abilities which I have been able to develop.

Nevertheless, the feeling remains with me that I have missed out on something. This feeling does not really disturb me provided I can maintain the impression that my work is really worthwhile.[8]

This final caveat nagged at him his whole life. When he felt confident about his research, he felt fine about life. But when work wasn't going well, everything else collapsed along with it.

Though women his own age were daunting, Roger had no trouble socializing with his younger sister, Shirley, and her joyous gang of playmates. They brought out a paternal tenderness in him. He made them toys and games, wood carvings and cardboard sculptures, and other whimsical, precisely mathematical delights.

"Every Christmas Eve, Roger would disappear into his room after supper," Shirley recalled. "I didn't know it at the time, but he'd stay up until three or four in the morning, and emerge with a new creature for me made out of table tennis balls and wire."

He made vividly painted caterpillars, ladybugs, and googly-eyed faces, all custom designed to delight Shirley.

In such a serious household, filled with serious men, Shirley and her friends uplifted the home with their noise and laughter. Even Lionel felt it, though he destroyed any evidence of his own moments of levity. Roger recalled,

> In the early days of tape recorders, there were various things we put on tapes—little things that people said. There was one occasion when my father completely got the giggles and wouldn't stop for ages. I think Shirley recorded this. This thing was amazing, that my father would completely succumb to the giggles. He just couldn't stop. Then he wiped it off. I thought this was stupid, and terrible. It was showing a side of himself he really had and evidently felt terribly ashamed of because he didn't allow this recording to be kept.[9]

One of Shirley's friends, a girl named Judith Daniels, had particular capacity to lighten the atmosphere. Judith was bold and warm-hearted, empathetic, and joyful. She always seemed to be rescuing a caterpillar or cat from peril, nurturing any creature in her care until it was ready to be set free again.

Judith's parents were divorced, and both had remarried. She and her sister Helen lived in London with their mother, Anne, and were close to all four of their parents. Judith's father came from a large Jewish family in Manchester, who were the opposite of the Penroses—down-to-earth, playful, and demonstrative in their affection for one another.

"Everybody adored Judith. Everybody remembered meeting her. She was very sparkly, had curly hair and twinkling eyes. She was a very beautiful child," said Helen. "She had a very quick mind. She took after our mother in that sense."

She also inherited her mother's laugh, a distinctive, unselfconscious snorting giggle that gave a lift even to the Penrose household's most dour inhabitants. "That laugh is worth a lot of money," said one visiting researcher, looking up from his perch as the sounds of Judith's joy filtered into Lionel's office.

Judith's father died of liver cancer when she was nine. Roger became a partial parental surrogate, entertaining and caring for her with the same tenderness he showed his sister. The whole family made the Daniels sisters welcome. Judith played piano. When Lionel wasn't around, she, Margaret, Roger, and Jonathan played together as a quartet.

Lionel's library was full of valuable old scholarly books and manuscripts. Roger loved to read these old texts, notably including a valuable manuscript of Johannes Kepler's. The sixteenth-century German astronomer had split his attention between decoding the physical world through geometry and playing with pure mathematics problems like tiling puzzles and other visual challenges.

Roger connected with the way Kepler blurred science and play. He pored over Kepler's idiosyncratic tilings of pentagons and dodecagons. They didn't fit together the way hexagons or squares did. Kepler used that to his advantage, creating complex, incomplete tiling patterns with their own irregular beauty. Roger appreciated how Kepler's tiling gaps and irregularities could be more interesting than standard tessellations.

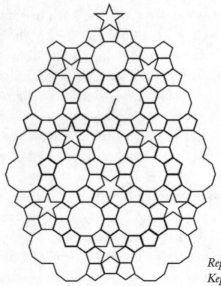

Reproduction of Johannes Kepler's pentagonal tiling.

At university, Roger relearned calculus, algebra, and geometry, taking them further and deeper than he ever had with his teachers or with Oliver and Lionel.

Numbers embodied so much beauty, power, and knowledge that he became convinced they had an external reality. They couldn't be merely a creation of humanity. "This is a very common feeling amongst mathematicians. The world of mathematics is out there, and it's a more like exploration than invention. They're discovering things that are already out there in nature in the Platonic world," he said.[10]

Numbers weren't the same as everyday physical objects—you couldn't hold "three" in your hand the way you could three tennis balls. But they were more than ideas. They were in the universe, available for people like Roger to explore their properties, appreciate their beauty, and put them to work.

The categories of numbers were like nested spiritual planes, expanding concentrically from mundane to exotic: natural "counting" numbers, whole numbers, integers, rational and irrational numbers, discrete and continuous numbers. The golden ratio, Euler's number, zero, pi, aleph null, googol. Every number had its own magic. Every formula and operation revealed new wonders. His whole life, Roger had picked up hints and echoes of this world-behind-the-world. Planets, jungles, fireplaces, and backsplashes were mere shadows cast on this reality by something more real, more essential.

He finally understood why the everyday world of impatient teachers and distant parents and confusing women so troubled him. He belonged to this other more vivid place, this Platonic mathematical realm where everything made elegant, beautiful sense.

One numerical concept captivated him above all others: the sublime poetry of complex numbers. "These numbers had an extraordinary magic and beauty about them," he said. "At that time, I had no special expectations to be a physicist. But I had taken a view that, 'Gosh, complex numbers are so amazing that it would be fantastic if the laws governing the workings of the universe were based on them.'"[11]

Complex numbers by definition mix reality and imagination. A complex number has two parts. The "real" part can be any standard number—whole, fractional, irrational, and so forth. The second, "imaginary" part is a multiple of i, which denotes the square root of −1. Is i actually any less real or more imaginary than any other number? In some ways yes: a negative number can't have a square root, so i can't exist. That's why René Descartes coined the term *imaginary number* in 1637.

But i is also incredibly useful. It's essential for solving cubic equations and other otherwise intractable mathematical problems. Graphs using real and imaginary numbers as function coordinates create completely unexpected curves and shapes, revealing whole new landscapes for mathematical exploration.

Roger would later discover imaginary numbers unexpectedly lurking in many equations of physics. The boundary between the real and the imaginary appeared to be permeable. Roger combined them with ease, using complex numbers to create vivid, unexpectedly simple mathematical descriptions of reality.

Oliver left London to do graduate work at Cambridge. He studied statistical mechanics, though when Roger visited, the brothers often returned to their shared interest in relativity.

Over lunch at the Kingswood Restaurant, their regular meeting spot, Roger puzzled out loud about how distant galaxies could possibly disappear from the observable universe.

"I don't know much about cosmology," Oliver told Roger. "But sitting at that table over there is a friend of mine who can tell you all about it."[12]

That friend was Dennis Sciama, a warm, generous, and well-connected Cambridge professor of physics. Dennis was happy to oblige Roger's curiosity. He pulled out paper and pencil and began drawing light cones and world lines and other diagrams to explain the universe's expansion and the inaccessibility of the most distant galaxies.

Dennis interspersed his impromptu science talk with stories about his recent encounters with Einstein. Dennis used to believe in a cosmological model Roger remembered from Fred Hoyle's lectures on the BBC, the *steady-state* model, which says the universe has always existed in its current form. He told Roger he had recently been talking to Einstein about steady-state theory, and Einstein had laughed him out of the room.

In the late 1920s, astronomer Edwin Hubble had observed how almost every galaxy in the observable universe was moving away from every other, like spots on an inflating balloon or raisins in a rising bread loaf. The only possible explanation was that the universe was expanding. Which meant it must have been much smaller in the past. Which implied the universe had a beginning.

The steady-state model could not be correct.

As their conversation continued, Dennis mentioned he worked with Fred Hoyle.

"Fred would be able to explain all of this better. I'll introduce you on another visit," he said.[13]

Roger, still an undergraduate student, couldn't believe his good fortune. The universe had done him a great service by placing him in the same restaurant as Dennis Sciama.

Back in London, Roger discovered new aspects of Lionel's blithe cruelty toward Margaret. With his sons less available to him, Lionel began inviting a young woman musician he knew around to the house.

"My father would play the piano and Selma would sing. She had a very good singing voice," Roger recalled. "I don't think there was any indication of infidelity, but he had a special relationship with her that I think trampled on Margaret."

Roger did not have the term *emotional affair* in his vocabulary and had trouble finding the right language to describe Lionel's mistreatment of his mother. "I don't think it was deliberate cruelty. I think it was insensitive cruelty. I think he was being insensitive to Margaret, but not deliberately trying to put her down," he said. He never spoke to his mother about it. He was much happier thinking about complex geometry and light cones.

He followed Oliver to Cambridge to begin graduate work in algebraic geometry. He seemed destined to become exactly what Lionel had feared: a pure mathematician. His interests, though, were evolving. When he looked up at the sky now, the physical world felt more transparent, the geometry more substantial. Reality started to look like a space-time diagram. "The different directions in the sky correspond to different null lines entering the point that is your eye. You're looking at your past light cone," he said. "If you want to describe the celestial sphere that we actually see, the neatest way of doing it is to think of it as a Riemann sphere."[14]

Bernhard Riemann, a nineteenth-century German mathematician, developed a coordinate system that uses real and imaginary numbers to demarcate vectors, angles, and points. The Riemann sphere has one point at infinity, located at its north pole. It was typically a tool for pure mathematicians, but Roger sensed the deep connections between complex numbers and the physical realm. He could practically see the sky overlayed with these coordinates.

4

THE IMPOSSIBLE
TRIANGLE

Just before 10:30 a.m. on the morning of September 2, 1954, Roger Penrose stood on the steps of the Royal Concertgebouw, Amsterdam's landmark concert hall. Built in the late 1800s, the building hulked over the wide and pleasant Van Baerlestraat Street, beyond which stretched Museumplein, an open, grassy public space bordered by high-profile cultural institutions.

The hall's imposing neoclassical columns rose up to the peaked roof, topped with a gilded sculpture of Apollo's lyre. Trolleys slid along out front, pausing to disgorge prominent mathematicians from over three dozen countries. Roger joined the flow as more than 2,000 of his contemporaries filed into the concert hall for the opening plenary of the International Congress of Mathematicians.[1]

Roger was still three years away from receiving his PhD. He was not speaking at the prestigious conference and had made the trip primarily to hear one of the world's most influential living mathematicians, Hermann Weyl.

Among many accomplishments, Weyl was famous for adapting complex geometry to physical phenomena like electromagnetic and gravitational fields—intellectual feats close to Roger's heart.

The overflow crowd filled the aisles and stairs of the main hall, which was decorated with flags from attendees' home countries. The prince of Denmark was abroad but sent greetings, congratulating attendees for choosing to "practice and promote a profession which, like few others, recognizes the unity of the human race."[2]

Many speakers addressed the responsibilities mathematicians bore in an increasingly technologized industrial world. Congress president Jan Arnoldus Schouten said in his address,

> I wish to draw your attention to a fact which was perhaps not so clear four years ago, but which is absolutely clear now: the place of mathematics in the world has changed entirely after the Second World War. During and after the war it became obvious to everyone that nearly all branches of modern society in war and in peace need a lot of mathematics of all kinds, from the simplest school arithmetics up to the highest developed theoretical parts. In fact, there is nowadays no big factory without its computing machines and no investigation involving...experiments or observations is possible without an elaborate application of modern statistics.[3]

Roger still doubted his decision to pursue mathematics. He'd proved his passion and his talent for the work, but his parents' anger and disappointment stayed with him. He was both comforted and intimidated to be walking among such prestigious company.

His PhD in algebraic geometry involved more algebra than he'd expected. He was more than comfortable with geometry—it energized and inspired him. Algebra, though, continued to daunt him. Schouten's remarks seemed designed to reassure him he didn't have to master both.

It is good to remember here a word of [famed French mathematician Henri] Poincaré's, who stated explicitly that in mathematics there are two kinds of mental acting, one above all occupied with logical deduction and the other guided by a more intuitive faculty for arranging or rearranging known facts in a new way satisfying some principle of aesthetics or of unification. Poincaré laid particular stress on the point that the choice of the method is by no means fixed by the matter treated and that it has nothing to do with the difference between analysis and geometry. There are famous analysts using largely the more intuitive faculty and famous geometers working as a rule with deductional methods.[4]

The grandeur of the event, the splendour of the hall, and the optimism of the talks all buoyed Roger's spirits.

Hermann Weyl rose to present the Fields Medal, one of the highest honours a mathematician can receive, to the 1954 winners, Kunihiko Kodaira and Jean-Pierre Serre. The prize is traditionally awarded every four years to extraordinary mathematicians under the age of forty. Weyl was nearly seventy that year and good-naturedly confessed both his admiration of and intimidation by new trends in mathematics. "I realize how difficult it is for a man of my age to keep abreast of the rapid development in methods, problems, and results which the young generation forces upon our old science," he told the crowd. "Be prepared then to have to listen now to a short lecture on cohomology, linear differential forms, faisceaux or sheaves on Kahler manifolds and complex line bundles."[5]

At every turn, the conference presented Roger with new reasons to feel emboldened—and new ways to feel overwhelmed. "It was intimidating, more than exciting. There were all these people doing things which I barely understood. It wasn't as though I got fired up by all this new stuff," he said.[6]

He arrived late for a packed lecture by newly minted Fields Medalist Jean-Pierre Serre on the homotopy groups of spheres, only to discover the subject was impenetrable and the lecture was in French. He edged right back out the door, left the building entirely, and descended the stairs just as a tram pulled up. Off stepped British mathematician and code breaker Shaun Wylie. In his hand, he carried a catalogue from an exhibition at the Municipal Museum of Amsterdam.[7]

Roger and Shaun recognized one another and chatted on the street, catching up on the conference and common friends back in England. Roger wasn't really listening. He was transfixed by the catalogue Wylie held. The picture on the cover depicted bucolic black-and-white checkerboard fields, separating two villages that were near mirror images of one another. Each had a church, windmill, treelined streets, and rows of houses. Though part of the same scene, the lefthand village was sunlit, while night had fallen on the right.

As Roger's eyes moved upward from the squares, the boundaries of the orderly fields at the bottom of the picture twisted and warped, morphing into the outlines of birds. On one side, the birds glowed white against the dark greys and blacks of the nocturnal scene. On the daylight side, they were black silhouettes on a white sky. Where the flocks met in the middle, black and white birds interlocked seamlessly—each becoming the background of the other.

Roger couldn't stop looking. He had seen many visual tricks and puzzles in his life, but nothing like this.

M. C. Escher's lithograph "Day and Night."

The catalogue came from an exhibition of Dutch artist M. C. Escher's lithographs, which had been organized in conjunction with the mathematicians' congress. Shaun pointed Roger toward the museum, which lay just across the Museumplein.

Escher's prints amazed Roger. He had never seen anything so clever, so beautiful, so surprising and delightful. They were realistic and impossible at the same time. Escher played with perceptions of up and down, foreground and background, scale and dimension. The images were a revelation, yet also deeply familiar, as though the artist had reached into Roger's own mind to pull out these ideas.

He stopped in front of a spectacular lithograph titled "Relativity." Three black-and-white staircases approximated an equilateral triangle. Platforms and smaller staircases connected the main flights, and arched doorways opened onto various idyllic gardens.

Abstract human figures moved through separate, overlapping worlds. Walls became floors, up became down, down became sideways.

M. C. Escher's "Relativity."

Beings that seem nearly to touch as they passed occupied disconnected realities.

Each small section of the picture made perfect sense. But as Roger tilted his head one way and the other, "Relativity" became increasingly irrational. Up, down, and around the staircases, paradox piled on paradox.

The image made powerful sense to him. Escher spoke in a visual mathematical language Roger understood fluently. The only other person he'd ever come close to sharing this language with was his father.

He spent the day awestruck and lost in the galleries. Interlocking birds turned into fish. Black and white humanoids stepped out of a two-dimensional image and marched around in circles to greet one another. Lizards climbed up out of the page and over a dodecahedron before slinking back into their flat world. Tesseracts and collapsing spheres, distorting mirrors, conformal stretching. A subtle running joke about how all these creatures and objects that seemed to force their way into higher dimensions only ever existed on flat pieces of paper.

He returned to the conference with his own copy of the catalogue, more energized by the exhibition than by the talks of his fellow mathematicians.

Back in England, he travelled from Cambridge to London at his first opportunity to share the catalogue with Lionel, who was as fascinated as Roger. In his twenties, Roger still felt the rush when Lionel looked at him with interest.

They began creating their own "impossible objects." They drew forests, roads, and streams, exchanging foreground and background and using perspective and two-dimensional trickery to create impossible scenes. They began with detailed, realistic images but quickly evolved toward simpler, more essentialised creations.

Each ultimately created one iconic stripped-down masterpiece, peeling away everything but the absolute essence of their illusions.

Roger borrowed the three-axis symmetry from "Relativity" to create the Penrose triangle. Each corner looked like a perfectly plausible sixty-degree angle formed of solid material. But something went wrong with the whole picture. Each of the drawing's twelve lines was perfectly straight, but the sides of the triangle seemed to twist—the inside became the outside, and vice versa. The angles seemed to oscillate between sixty and ninety degrees. The Penrose triangle was irreconcilable with itself.[8]

The Penrose triangle.

Lionel, inspired by Roger's triangle and Escher's stairs, developed the Penrose staircase, four flights of stairs connecting the top edges of a seemingly rectangular tower. A person walking along those stairs would ascend (or descend) forever. The staircase had no beginning, end, top, or bottom.

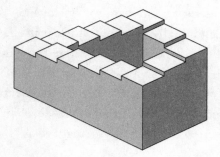

The Penrose staircase.

While Lionel and Roger collaborated on impossible objects, Dennis Sciama worked to advance Roger's scientific career. He was convinced Roger's brilliance and curiosity were better suited to physics than pure mathematics. "Dennis was a very great promoter of friendships—getting people together who he thought might be of interest and importance to each other. He had a very wide collection of friends. He was a very good at talking to people on all sorts of topics, but primarily physics. That's what he was really interested in. He developed this special friendship with me. He was determined to teach me physics. Somehow, he thought this was worthwhile," Roger said. "We used to go to Stratford to see Shakespeare plays. Dennis was very keen on that. We might stop off at Oxford on the way to see someone Dennis thought I should meet."

Dennis had all of Lionel's astuteness and none of his coldness. He found joy and science everywhere he looked. He drove at breakneck speed along twisting country roads, gleefully expounding on the cosmological physics at work. Every time Roger was crushed into his door on a sharp turn or lurched forward at a sudden stop, Dennis identified it as a manifestation of Mach's principle, a hypothesis that the inertial forces of a body in non-uniform motion are determined by the quantity and distribution of all matter in the universe. Roger recalled,

> He was very keen that our local inertia should be determined by the distant stars. I had these conversations with Dennis, where we imagined the stars being removed one by one. You remove the Earth as well, and finally, you get to the car itself. Would the car determine the inertia in the universe? He thought it would. So I would then dig down—suppose you just have one or two particles. How do you determine the angular momentum in relation to other particles? And out of this I later developed a "theory of quantized directions," which was formalized as "spin networks."

Physicists later abandoned Mach's hypothesis in favour of other characterizations of inertia, but starting in the 1970s, Roger's spin networks—a diagram system for representing states and interactions of quantum particles and fields—would become an essential component of modern theoretical physics.

In Dennis's push for Roger to break away from pure mathematics, he directed him toward three life-changing lecturers: Hermann Bondi on general relativity, Paul Dirac on quantum theory, and S. W. P. Steen on mathematical logic. "In different ways, they were each absolutely beautiful," Roger recalled. "They influenced me more than any pure mathematics course."[9]

Bondi taught general relativity on a much deeper level than Roger had ever known it. The geometry he had played with since he was a child now opened his eyes to whole new realms of scientific discovery.

Dirac's lectures on quantum mechanics affected him just as profoundly. At smaller and smaller scales every physical property seemed to disappear: colour, temperature, pressure, conductivity, elasticity. These characteristics gave way to deeper, more fundamental properties. Ultimately, the only things that remained were numbers and equations. Mathematics underlay it all.

At a pivotal moment, Roger tuned out and missed Dirac's somewhat famous explanation of how tiny quantum systems transformed into macroscopic classical systems. "People told me he takes out a piece of chalk and breaks it in two to illustrate particles being in two places at once. Then he gives his explanation. I probably heard it, but I didn't listen to it. I was staring out the window or something. From then on, I worried about this question. Now, it's probably just as well I didn't catch his interpretation that would have been some kind of waffle about why big things require more energy. That's part of the story, but it's not the real story," he recalled.[10]

He suspected Dirac's reasoning would have corrupted his own exploration of the relationship between quantum and classical physics. Every interpretation of quantum mechanics felt like waffle to him. He dismissed anyone pretending to have an explanation when quantum theory was still so incomplete. He kept his thoughts to himself, sensing his scepticism was not welcome among experts in the field.

Roger took more notice when Dirac diverged from his standard lecture material to talk about "two-component spinors," which used complex numbers to describe quantum properties like the spins of subatomic particles. Dirac prodded Roger's vague but passionate hope that complex numbers might yet prove essential to understanding reality.

From Steen, Roger learned about Kurt Gödel's incompleteness theorems. Gödel's 1931 theorem upended mathematics, showing how every logical system generates statements that are true but that cannot be proven within that system. At the most basic level, Gödel statements took the form "This statement is not provable within this system." If it *could* be proven true, that would mean also it *wasn't* true—a paradox. If it *couldn't* be proven true, the statement was true but was obviously unprovable.

Demonstrating this limitation of mathematical systems led Gödel to the related question of how mathematicians gained the mysterious power to transcend it. A Gödel statement couldn't be proven true, but any Roger, Kurt, or Dennis could *see* that it was true. If they weren't using logic to determine its truth, what exactly were they using?

Gödel showed that understanding transcends any system of rules. In other words, it transcends computation. I was most impressed with this argument. You try to prove things in mathematics using a framework of rules. Gödel made a statement that said, "I am unprovable by these rules." Now, that's a clever thing to do. It's very clever because it's telling you, "Here is a statement that you know to be true, but you can't prove to be true using the rules of the system." Your trust in the rules and your understanding of why the rules work transcend the rules. And that's stunning to see. Absolutely stunning.[11]

Roger had been watching brilliant people solve puzzles, make discoveries, and play games his whole life. In Gödel's theorem, he caught a glimpse of something he had long suspected: a missing element in explanations of human intelligence, understanding, and consciousness.

As a lifelong atheist, he knew the mysterious, non-mathematical nature of human understanding must derive from physics. But no current physical theory could explain it. "The universe doesn't know what it's doing. But somehow evolving consciousness is hugely effective in natural selection. At what stage did that happen? How does that happen? At what level? Those are fantastic questions," he said.[12]

At Cambridge, he still brought his Perspex tetrahedron puzzle to social events, finding a quiet corner in which to set it up. He established some lifelong friendships, including with Ivor Robinson, a physicist he met through Dennis. Ivor was a few years older than Roger and had none of his shyness. From a Liverpudlian Jewish family, Ivor loved to expound on science, politics, music, and most other subjects. Roger loved to listen.

For the final year of his PhD programme, Roger lived with his parents' friends Max and Lyn Newman, who had a house in Comberton, a village of thatched cottages and quiet roads biking distance from Cambridge. He was happy being away from his parents and got on well with Max, who was much more relaxed than Lionel in most things. The one notable exception was his piano. He could not bear to share his instrument with others. He forbade Roger—or anyone else—to play it.

"If Max was there, I would not even be allowed to touch the piano. He was a very good pianist. He even visited Princeton and played the piano with Einstein playing the violin. He said the trouble with Einstein was that he couldn't keep time properly."[13]

Max held the Fielden Chair of Pure Mathematics at Manchester University and was often away for the day or overnight. When he had the house to himself, Roger stole into the sitting room, slipped back the fallboard of Max's grand piano, and played for hours at a time. He was self-conscious about his musical skills and preferred playing when nobody else was around to hear him.

He knew few pieces by heart and improvised in the style of J. S. Bach or W. A. Mozart, picking out simple melodies and extemporizing variations, harmonies, and counterpoint. He fantasized about composing his own music for publication and performance but never took it beyond these furtive sessions. When he finished playing, he fastidiously wiped his fingerprints off the keys, ensuring Max was none the wiser.

In September 1956, Roger submitted his PhD thesis, "Tensor Methods in Algebraic Geometry." Tensors are algebraic objects that describe the properties of physical objects in simple numerical or graphical terms. Tensors use *vector space* to quantify relationships between physical phenomena like stress, elasticity, force, and velocity, as well as size, shape, and many other characteristics. Roger sought more generalized, universal applications for tensors. "The author's aim...is to develop a general theory of abstract tensors, without reference to the 'components' of the tensors in a 'frame of reference.' The objects considered are of a very general type (much more general, in fact, than are ever needed in his applications) and the author's evident ambition is to have a theory wide enough in its scope to describe almost any conceivable geometric object," wrote J. A. Todd, a renowned geometer and one of Roger's thesis reviewers, in his report to the committee. All the reviewers liberally sprinkled their reports with words like "original," "ambitious," and "by no means light reading, especially after the first fifty or sixty pages."

Roger filled more than 200 pages with neologisms and an alphabet of ornate quasi-letters he designed himself to represent terms

A section of Roger Penrose's PhD thesis.

and concepts crucial to his arguments. The reviewers were both impressed and overwhelmed by Roger's relentless intellectual density. Todd wrote,

> The details of this work are extremely complicated, and in order to get a reasonably concise notation the author has had to introduce a number of abbreviations and special symbols, which make checking details of the proofs a very difficult matter, and the extreme generality of the author's approach spares neither himself nor the reader. I have not attempted to check all the details, but I feel confident that they are correct. The author's style is far from graceful, but the material is sufficiently intractable for this to be excusable. The ambitious nature of his very general approach commands respect. The originality of his approach, and the great formal skill shown in handling very difficult material, make this a dissertation quite out of the usual class....Few people will have the courage to read this work...but of the high originality in its conception and of the masterly technique shown in its execution, there cannot be any doubt. The degree of Ph. D. is, of course, amply earned, I would not like to estimate how many times over.[14]

On March 4, 1957, the Degree Committee of Cambridge's Faculty of Mathematics unanimously awarded Roger a PhD. He returned to his parents' house in London, having secured his first academic position as an assistant lecturer at Bedford College.

Bedford, founded in 1849, was the United Kingdom's first higher education institution exclusively for women. "Me having very little experience with women, it was sort of an unusual experience to teach these classes full of girls," he said.[15] His enthusiasm for mathematics outshone his awkwardness. He was not much older than his students, looked young for his age, and had an informal teaching style that made him a popular prof.

Only his shyness, a vague sense of propriety, and the indifference of the era kept him from scandal.

You know, you shouldn't have a relationship with a student. But when the year was finished, when exams were finished and everything, there was a ball. I took one of my students to the ball. I was actually more interested in her good friend, which is pretty stupid because she was probably the worst one in the class. The one I did take had a distinguished father who was a member of the Royal Society. The one I didn't take, I remember playing tennis with her on a court after the ball in the middle of the night until dawn broke. I never had much of a relationship with her. I was just a little besotted at the time.[16]

The relationship never progressed beyond that night, playing match after match, as Earth slowly turned them back toward the sun.

Living again with Lionel and Margaret, he rediscovered the joy of entertaining his sister and her twelve-year-old friends. He paused each day on his way to the college to chat with Shirley, Judith, Helen, and the other girls waiting together for their school bus. He carried on to the college uplifted and confident.

On weekends, he took them to visit the London Zoo or out on pedal boats in the park. "It was in this wonderful location in the middle of Regent's Park. There's a lake system you could go through on a boat. I took Judith and Helen out on boat rides, which was very lovely."

The Daniels girls also were regular visitors at Thorington. They didn't mind the warped, uncarpeted floors and drafty bedrooms and were wowed by the luxury of a private tennis court. Roger found it strange and liberating to have young friends, free of the intensity and judgement that permeated his adult world. Judith exuded joy and laughter and openness—all the things Roger's family lacked.

Roger spent only a year at Bedford, after which he returned to Cambridge for a postdoctoral research fellowship at St John's College. He was back among Paul Dirac, Dennis Sciama, Ivor Robinson, and other men he admired. He didn't yet see himself as one of them. He had yet to fully prove himself as a mathematician, let alone as a physicist.

He no longer relied on Max and Lyn's hospitality. The fellowship came with prestigious lodging in historic Shrewsbury Tower, perched over a high archway connecting two of the college's campus courtyards. "It was nice. Romantic in some sort of way I suppose," Roger said. "I guess I felt a little lonely when I went back to Cambridge."[17]

Dennis Sciama's office at Cambridge was a social and intellectual hub, where men of science came and went, trading gossip and new theories about the state of the universe. Roger was surprised on one such visit to find a young woman working at a desk, surrounded by illustrators' tools. She had shoulder-length light brown hair and dressed less formally than the men in their jackets, crisp shirts, and narrow ties. He found something comforting about her rumpled clothes and half-hearted hair and makeup.

Dennis introduced him to Joan Wedge, an American expat who was helping with some illustrations for his upcoming book. Joan spoke with a Michigander accent. She was abrupt and informal, friendly and unguarded—very unlike anyone Roger was accustomed to speaking with at the university.

Roger took an interest in her illustrations, which were destined for Dennis's short treatise *The Unity of the Universe*. They appealed to his visual approach to cosmology. Asserting his now professional interest in physics, he offered to help her with the science behind her illustrations. Soon he was visiting the office more to talk to Joan than to Dennis. Science illustration became the bridge that brought them together.

She was just a few months older than him but came from an utterly different background. Joan's paternal grandparents had been wealthy, but they "lost it all" when she was a little girl growing up in Pennsylvania. "I was sort of all over the class system in America," she recalled.[18]

Joan and her parents moved to a working-class neighbourhood in Detroit, Michigan, where her father sold life insurance. He had trouble finding steady employment.

Joan studied art and teaching at the University of Michigan, though she downplayed her education. "I majored in boys," she said. "I had an absolutely wonderful time. I never studied. I just picked guys who were good at whatever I was taking. They would coach me, and I did all right."[19]

In 1952, she met British astrophysicist Bernard Pagel, who was in Detroit on a Fulbright scholarship. "Bernard...was so much better educated than the others and he had a point of view," she recalled. "I remember once he was explaining to me why capital punishment was not a good idea from a societal point of view. I had never heard anybody talking about it like that."[20]

When Bernard wrapped up his year in Michigan, Joan planned to follow him back to England with intent to marry. She spent several months working at a drafting company, saving money for a boat ticket. On June 29, 1954, she stowed her single trunk on board the *Arosa Star* and set off from New York.

She and Bernard broke up soon after. Joan stayed in England, found a job teaching at the American School in London, and started dating Dennis Sciama. The American School had small classes, paid well, and gave her the flexibility to spend half her week in Cambridge. Every Wednesday afternoon, she loaded her bicycle onto a train at St Pancras station and travelled up to work on Dennis's book. Their relationship ended before the book was complete.

Roger had never gotten to know a woman his own age before. He had never met anyone with Joan's openness and sense of adventure. She was messy and chaotic and didn't seem to care what people thought of her. "One of the things that drew me and Roger to each other was that we were both very untidy," she said. She was attracted to smart men. The twinkle in Roger's eye and his winning smile drew her only because they were enhanced by his overflowing cleverness.

Roger was mystified and gratified by Joan's enthusiasm. They talked and worked late into the night. Eventually her boldness overcame his awkwardness. At age twenty-six, Roger found himself in the first intimate relationship of his life. "It was in that room in Shrewsbury Tower where I remember really getting to know Joan. That was really the first time I had a sexual relationship with anybody. That probably makes more of an impression on one than it should. She kind of made a beeline for me," he said.[21]

They shared meals at the Kingswood Restaurant and talked and walked along the River Cam, over Cambridge's stone bridges, and through grassy quads. After only a few months, Joan began talking about marriage—something he had never considered for himself.

He was abstractly aware that Joan suffered periodic bouts of depression. When she was sad or angry, neither of them knew what to do about it. He assumed his rock-solid rationality would eventually trump whatever troubled her. He remembered,

> I had this incredibly bloated opinion of my myself thinking that she obviously had psychological problems, and getting married was going to sort her out and everything would be fine and lovely. That was the feeling I had, completely overrating or misunderstanding what relationships are like and thinking that somehow the problems she had would dissolve away with the nice happy life that she was going to have with me. I'd never really had a proper relationship with any woman before that. You have to bear in mind, I was terribly shy with women. I knew some. I had a crush on one or two from time to time. But I never really got to know any woman closely. This was the first time I ever did. So probably I was more susceptible than I should have been. It was really very unsuitable. It became more obvious later, but it was a great mistake.[22]

He proposed to Joan in 1957.

If Lionel and Margaret had been unhappy about Roger's choice of mathematics over medicine, they were livid about his plans to marry Joan. "Neither my mother nor my father approved of this at all," Roger recalled. "They thought it was completely the wrong thing for me to get attached to this woman."[23]

Not long after Roger announced their engagement, Lionel paid a visit to Joan under the pretence of wanting to get to know her better.

His true intentions quickly became clear. "Lionel was enormously rude," Joan recalled. "He said he wanted to come over to see what I was really like. I remember he started off by telling me that he heard my father was mentally very disturbed and probably not deserving of contributing to the Penrose gene pool. And then told me how untidy I was, and then told me he didn't like me very much. And then he left."[24]

In his papers and books, Lionel argued against eugenics, insisting the word be removed from his title. But he went to extremes to keep the Wedge line from contaminating the Penroses.

When he failed to put Joan off Roger, he shifted to putting Roger off Joan. "My father went on a very peculiar route. He considered her so odd because she had some strange movements. He said she had the early symptoms of the disease Huntington's chorea. His judgment was completely distorted by the emotional part of this whole thing. He thought I was throwing my life away, which may well have been right, but being young and thinking too much of oneself, I thought I knew better," Roger said.[25]

Lionel seized on Joan's benign physical tics and enlisted a colleague to put her through a battery of tests for Huntington's, a debilitating condition that robs people of the ability to eat, move, and speak.

His colleague dismissed the idea out of hand. Roger was more puzzled by his father than angry. "I mean, this was really odd, because he was somebody who had really good instincts, keen instincts with people and I used to trust him very much. And this was ridiculous. This was a completely wrong diagnosis. Why did he invent the idea of this disease to explain her strange movements? My father was going off the rails, but reasonably, because he was so upset that I was getting married to this woman," he said.[26]

When his genetic gambits failed, Lionel tried, as a last resort, talking to Roger. He invited his son for a walk in the woods. Rather than to play Kriegspiel chess, though, Lionel used this walk to tell Roger a story from the days before he met Margaret, when he lived in Austria with Max Newman.

"He talked about some woman he had known in Vienna," Roger recalled. "She must have been very attractive, I suppose, because he claimed Max and he were both very attracted to her. And then he realized that she was slightly disturbed and that he wanted to have nothing to do with her."[27]

Lionel waited for Roger to see his point—in vain.

"Since he was so indirect, it was unclear what he was talking about. I just thought this was him being unreasonable," Roger said. "The problem was that my father couldn't express what he meant about it."[28]

Roger's picture of why he married Joan was like a Penrose triangle: consistent in discrete parts but self-contradictory when combined. He was passive and helpless, swept along by Joan's determination to marry him. But he was also stubborn and wilful, defying his parents and defending his right to decide for himself what to do with his life. And underneath it all—her desire, his parents' disapproval, his own muddled intentions—was the indifferent cascade of physical causes and effects that governed every particle in the universe. His body and brain were made of matter, subject to the laws of relativity and quantum mechanics he had studied at Cambridge. Did he *find*

Margaret and Lionel, newlyweds Joan and Roger, and Oliver and his wife, also named Joan.

himself married to Joan, or did he *decide* to marry her? Roger leaned toward marriage having happened to him.

"Space-time is out there, and I am exploring it with my life," he had told his Latin teacher. A relativistic universe had no room for free will.

In 1958, he and Joan married in the university courtyard below Shrewsbury Tower, with Lionel and Margaret grudgingly in attendance. Cambridge remained their home for a few more months while Roger sorted out the next chapter of his career and Joan prepared for life as a wife and mother.

Later that year, Lionel and Roger coauthored a short paper in the *British Journal of Psychology* (where Lionel was friendly with the editor), exploring the mental mechanisms behind their Escher-inspired impossible objects. "Each individual part is acceptable as a representation of an object normally situated in three-dimensional space; and yet, owing to false connexions of the parts, acceptance of the whole figure on this basis leads to the illusory effect of an impossible structure," they wrote.[29]

Shrouding the Penrose triangle in abstruse academic language did not disguise its delicious simplicity: a child could sketch an impossible tribar, and yet its contradictions fascinated mathematicians and artists. This paper, coming so early in Roger's career, helped define him as a master of paradox and a creative thinker with appeal beyond the academic world.

Lionel mailed the paper to M. C. Escher. "The triangle really belongs to my son, Roger, and the staircase was my contribution. We have a number of ideas which we wish you could illustrate because we are not clever enough to do this ourselves," he wrote.[30]

Roger and Lionel also collaborated on a second publication that year, an eclectic collection of geometric puzzles for the Christmas issue of *New Scientist*, "to provide mental stimulation during the academic vacation." They challenged readers to weave kite strings over and under bridges, cover an infinite plane with strange-shaped tiles, and design a billiard table "with the property that there are two regions, A and B...such that it is impossible to hit a ball from any point inside A to one inside B."[31]

Their combined creativity seemed to overpower disagreements over Roger's wife and career. Roger didn't consciously choose to forgive his father. He just focused on other, more pleasant things.

As Roger's interests shifted from pure mathematics to mathematical physics, he sensed he needed something to help him stand out, to prove he had the chops for the job. "What do I know about general relativity that other people don't know?" he asked himself. "How could I get an angle on the subject that would be different from other people's? Ah! Two-component spinors!"

Roger realized he could use the spinors he'd learned about from Paul Dirac to rewrite Albert Einstein's equations in a new and simpler way. "It just worked so beautifully," he said.[32]

In June 1958, Roger attended the International Conference on General Relativity and Gravitation in Royamount, France. Dennis Sciama was an invited speaker. Roger intended to just listen, but at the last minute, Dennis offered to give half of his allotted time to Roger. His spontaneous generosity put Roger in front of an esteemed community of physicists from around the world, where he impressed them with his talk titled "General Relativity in Spinor Form." "We know from Dirac's theory of the electron that spinors have a fundamental importance to physics in the small, perhaps more so than vectors. I hope to indicate that this may also be true for the large-scale phenomena of gravitation," he told the gathered scientists.[33] His long-held, private wish that complex numbers might underly the whole physical universe was now a full-fledged thesis.

Roger's spinor approach revealed insights others had missed. He remembered George Gamow's charming scene of Mr Tompkins witnessing a bicycle contracting and compacting as it accelerated toward the speed of light. Roger knew now that Gamow had it wrong. While space would contract around a fast-moving object, the object would not appear flattened as a result. He used a sphere rather than a bicycle in his proof, but the theory applied to both. "It would be natural to assume that, according to the special theory of relativity, an object moving with a speed comparable with that of light should *appear* to be flattened in the direction of motion," he wrote in *The Mathematical Proceedings of the Cambridge Philosophical Society*. "It will be shown here, however, that this is by no means generally the case."[34]

The talk Dennis facilitated for him in France and the paper "The Apparent Shape of a Relativistically Moving Sphere" were major milestones. The triangle paper brought him closer to his father, and the sphere paper brought him closer to finding his own direction and purpose. He discovered the existence of people who admired his ideas and took no issue with his love of mathematics.

He cultivated a second family tree: an intellectual lineage separate from the Penroses, Peckovers, and Leathes. He inherited ideas passed down through generations of scientists: Johannes Kepler, Isaac Newton, Albert Einstein, Paul Dirac, Ivor Robinson, Fred Hoyle, and Dennis Sciama. Another branch of this tree held his mathematical forebears: Euclid, Hermann Weyl, Hermann Minkowski, Karl Friedrich Gauss, René Descartes, and a pantheon of others.

These mythic titans became role models he could imagine emulating. One day he might even take his place alongside them.

The prospect was exhilarating and intimidating. How could he measure up to the most accomplished scientists in history when his own brother easily outshone him academically? When he couldn't beat his younger brother at Rock Paper Scissors, let alone chess? When his disappointed parents lamented his pursuit of mathematics and physics?

He was chasing the biggest questions in physics and philosophy: Why is there something rather than nothing? Why are things the way they are? How can we know what the universe is *really* like? His father had taught him there was no inherent difference between this kind of high-minded conundrum and the riddles and games that cluttered the tables and conversations of his childhood. They were all just puzzles, waiting to be solved. He was good at puzzles. Just maybe he could solve something profound.

5

BLINK BONNIE

In 1959, Roger left Cambridge for a two-year research fellowship funded by the North Atlantic Treaty Organization (NATO), which sent him to Princeton for a year, followed by shorter stays at Syracuse and Cornell universities.

Roger and Joan travelled by boat to New York and on to Joan's parents' home in Detroit.

The Wedges lived in a small, run-down house on a short residential strip within a largely industrial neighbourhood. A freeway was under construction just a stone's throw up the street.[1] Roger did not care for the house or the location. His interactions with Joan were strained, and he had trouble making conversation with her parents, Utley and Helen. They knew very little about mathematical physics, and he knew very little about anything else. He wanted to get away as quickly as possible.

Lionel still owned the cottage across the Canadian border in Bayfield, Ontario. He invited Roger and Joan to stay at Blink Bonnie for as long and as often as they would like and even offered to make the cottage a belated wedding gift from him and Margaret.[2]

Roger and Joan bought a car and drove the three hours up along the eastern edge of Lake Huron to survey Blink Bonnie. Lionel had hired a local couple to look after the place when the Penroses weren't around. The couple's son, Charlie Kalbfleisch, met Roger and Joan to give them a key and show them around. The Kalbfleisches had put a new aluminium roof on the cottage, but several windows had leaked the previous winter, and the whole cottage smelled musty and dank. Wallpaper peeled in the bedrooms and living room, and the smell coming out of the kitchen sink drain was overpoweringly awful. The mildewy basement was overrun with insects and mice. The two-burner hotplate was the one Margaret had used years before. The cottage still had no stove or refrigerator. The wiring and plumbing worked, but they had to scramble under the building to prime a pump in order to run water. The well needed to be cleaned and lined. The backyard was covered in manure left by passing horses from nearby farms.[3]

He and Joan still managed to make Blink Bonnie a temporary home base. "The place seems liveable in, but probably not too comfortable," Roger reported to Lionel. "We very much appreciate the offer of Blink Bonnie as a wedding gift, but it is difficult to see that we would be able to make use of it or the land very much in the foreseeable future."[4]

In Detroit, they bought a huge trailer for their small Saab. They hauled a stove, refrigerator, furniture, and household goods across the border and up the rough highway to Blink Bonnie. They retiled the kitchen floor, tore off the surviving wallpaper and added fresh paint, turned and levelled the soil in the yard, and planted new grass and hedges. They improved the drainage and dried out the basement. The kitchen sink still emitted a reek, and the well still needed work, but the cottage started to feel homier. Roger reconsidered Lionel's offer and, in 1960, took over ownership of the cottage.[5]

Joan was about seven months pregnant when Lionel and Margaret shipped them a portable crib. She responded gratefully, attempting to rise above Lionel's previous cruelty. "The Karry Kot arrived in good shape and is now all set up and ready for action, I'm sure it will be very suitable for our nomadic life. As a bonus it is exactly the right size for carriage size fitted sheets they sell over here. Thank you very much for giving it to us—it will be Charley's only piece of furniture for quite a while," she wrote.[6]

"Charley" was a placeholder—they didn't know the baby's sex. With Roger's career on track, tentative peace with Lionel and Margaret, and improvements underway at Blink Bonnie, Joan optimistically described herself as "a lady in waiting," excited to welcome their son or daughter.

In Princeton, they rented the attic of a house belonging to the administrative assistant of Princeton physics professor John Wheeler. John was a vocal supporter of Roger's spinor work. (He was also the kind of physicist Roger's parents had warned him about: in the early 1950s, John had run Matterhorn B, the research project behind the hydrogen bomb.)

The attic was small and close but affordable and near campus. Joan had it to herself during the day, waiting for Charley to arrive with only the radio for company. American politics fascinated and horrified her. She wrote,

> We are both for Kennedy and I haven't a clue as to whether we hold the majority opinion over here. It is quite clear at this point that Kennedy is a much tougher, more self-assured cookie than Nixon. Although I don't think Nixon is honest or has much of a point of view of his own, I think the more fundamental and interesting qualities about him are his inadequacies as a grownup. Although I'm sure he is very bright with a tremendous amount of natural ability, his every move and word seems to be part of a performance and from what I have seen he exists only as the image he wants to see reflected in the eyes of the voter, or the president, or a politician....I bet anything the Real Nixon is a stunted little 7-year-old (or something like that) who never got a chance to grow up.[7]

While Joan followed the presidential race and Roger chalked out the passage of time on his office blackboard, Charley's due date came and went. They discussed inducing labour, but their doctor recommended against it. Roger didn't register why. They accepted his advice.

Nervous days turned into torturous weeks. Joan was no longer a lady in waiting but a terrified expectant mother. By the time they knew what was happening, it was already too late.

Their daughter was stillborn.

Their loss compounded with questions about what might have been. What if they had induced labour two weeks earlier? What if they had never come to America and instead relied on England's medical system? These questions were as useless as they were inevitable. The past was untouchable. The present was unbearable. The future was unthinkable.

Joan was crushed beyond comfort. She didn't leave behind a journal or letter describing her loss, but it affected her for many years, dulling joy and deepening sorrow.

Roger realized he had been unconsciously hoping for a daughter. Some of his happiest memories came from looking after Shirley, Judith, and Helen. He too developed a permanent ache for the daughter he would never have.

Unlike Joan, though, Roger was able to compartmentalize and get back to daily life. He reacted to grief by escaping it, throwing himself into Princeton's welcoming community of physicists. His time there coincided with a resurgent interest in relativity. In the 1940s and 1950s, relativity had taken a backseat to nuclear and particle physics.[8] John Wheeler built and mentored a growing community of brilliant young researchers who rang in a new "golden age of relativity."

Roger scrambled to figure out who was who and how he might fit in.

"We have some interesting people here," John assured him. "Let me introduce you to Dickie and Reggie."

Roger found Wheeler's use of diminutives oddly informal for such an august institution until he figured out he was referring to American cosmologist Robert Dicke and Italian physicist Tullio Regge.

Roger walked into Tullio's office to find the Italian geometer sitting in an odd chair shaped like a distorted torus.

"Ah, that's a Dupin cyclide!" he said with instant recognition.

"Not many people know that," Tullio replied, impressed.[9]

From that moment on, Roger felt he belonged. At Princeton, he didn't have to apologize for his enthusiasm or explain why he approached physics through geometry. He felt respected, valued, and even liked. His social life revolved around the physics department, which meant Joan's did as well.

Joan remained inconsolable, which made Roger uncomfortable. He still expected that being married to him, especially with his career going so astoundingly well, would rub off on Joan, and she'd be happy as well. When this continued not to happen, he avoided his puzzlement and disappointment by disappearing even further into his work. The universe was very big. His and Joan's problems felt small in comparison. He carried on with lectures and research, discounting his own sadness, as well as Joan's, as cosmologically inconsequential.

Roger's spinor-based relativity continued to generate talks and papers, enhance his reputation, and give him a sense of purpose. "The beautiful complete symmetry of the 'gravitational' spinor and its attendant properties was mainly what enticed me irrevocably into the subject of general relativity," he later recalled in his collected works.[10]

Empty space, free from the influence of gravity and electromagnetism, still had a shape, a curvature that could be quantified, calculated, diagrammed, tested, and understood. Roger fled deeper and deeper into that orderly mathematical world, whose beauty made his everyday life with Joan seem all the uglier.

Doubts troubled him. Lionel and Margaret had been dead set against his marrying Joan. He hadn't forgotten Lionel's mean-spiritedness or the desperate Huntington's gambit, but he wondered if his father's underlying motivation had been well founded.

He also began crafting another narrative: he started to believe Joan had tricked him into marrying her. He reimagined his shyness as lack of interest and her affection as coercive. He began to tell himself he hadn't had any say at all in his marriage. It had been forced upon him. And now he felt powerless, capable of neither mending his relationship nor ending it.

The couple left Princeton in winter 1961 for upstate New York. At Syracuse University, Roger found another community of talented and inspiring scientists who readily welcomed him and Joan. Joan remembered,

> I felt when we were there that Roger was happier because there was a source of warmth. You felt like you were part of a warm family. And I was accepted as well. There was Roger and me, and there was Ivor [Robinson] who was on his own. And there was Engelbert [Schücking]. And the Trautmans. Every night we went out for dinner at each other's houses. We called it "the kibbutz." They were so entertaining, Engelbert and Ivor. They were both larger-than-life characters. Who needed television? It was like being part of a big family.

Their meals had all the rowdy joy Roger remembered from his parents' dinner table, with none of Lionel's brutal scrutiny. Over wine and food, Ivor and Roger renewed the friendship they had begun in Cambridge years earlier. Like Roger, Andrzej Trautman came from a family of artists. He wrote his PhD thesis on gravitational radiation at a time when many physicists were still unsure it even existed.

Gravity was unlike electromagnetism or the strong and weak nuclear forces. It wasn't a force at all but a relativistic a warping of space-time. Relativity said spinning, moving objects might create *gravitational waves*—ripples in space-time moving across the universe at the speed of light. These waves would carry energy away from the source, which would have a huge impact on modelling of large and fast-moving objects.

Trautman and his partner, Róża Michalska (they wouldn't marry for another year), had been graduate students together in Poland, studying under Leopold Infeld. Infeld, who had worked closely with

Albert Einstein, doubted the existence of gravitational waves. Róża and Andrzej both believed (correctly) that gravitational waves existed and were developing mathematical solutions to prove it.

Andrzej and Róża lived in an apartment directly above Ivor. Their building became the hub for kibbutz dinners. Visiting scientists and mathematicians often joined the core group for food, physics, and fraternity. "[We talked about] politics and science and politics and Jewish problems and so on. It was very pleasant," Andrzej recalled in a 2019 conversation with American physicist Donal Salisbury.

Andrzej and Roger agreed on many fronts but disagreed about spinors. "I differed with Roger because I recognized from very early on the validity of the mathematicians' insistence on considering defining spinors in manifolds in terms of spin structure.... He did not accept spin structures. Roger is a genius, and it is interesting here: He is a mathematician by education... but interestingly enough, he had not had a very good education in modern mathematics."

Disagreements only enriched the conversations. This group, gathered from across Europe, with contradicting ideas and theories, all spoke the same language. It was a joy to sit down with them each night, to listen and argue and belong.

Joan was delighted just to be a fly on the wall. "When you've got all these people with this absolute love of their work, they live, eat, and breathe it. And that is a wonderful thing. I mean, I wasn't a part of it. But I very much enjoyed watching from the sidelines," she said.[11]

Engelbert Schücking, Roger's officemate, had studied with Werner Heisenberg in Göttingen and shared Roger's interest in geometric relativity. Engelbert was just beginning his illustrious career as an astrophysicist, relativist, and cosmologist. Roger quickly developed great admiration for Schücking, who seemed to understand many mystifying areas of physics.

"Engelbert had more influence on me scientifically than anybody else. There were two things in particular. In Syracuse, he told me about the importance of conformal invariance in Maxwell's equations.

I had some feeling for conformal geometry already, but it was mainly spatial, not for space-time."

In conformal geometry, size and distance have no meaning as long as angles are preserved. Light cones are a simple example of *conformal invariance*. If you change the time and distance units from, say, light seconds and seconds to light years and years, the cone will look the same and obey the same mathematical rules.

Engelbert guided Roger through the conformal invariance of the equations of James Clerk Maxwell, which form the foundations of electromagnetism and optics. The Maxwell equations deal in photons, whose masslessness make possible their conformally invariant behaviour. Roger soon began speculating about how to generalize conformal geometry to "remove the ruler" from physics equations altogether, creating a single set of laws for the universe at every scale.

Engelbert's second revelation helped Roger develop his understand of quantum mechanics, which had always confused him. "He explained to me what he regarded as the key thing about quantum field theory. Nobody else had said this to me. I'd never seen it anywhere else before. He said the fundamental thing in quantum field theory was the splitting of field amplitudes into positive and negative frequencies. You take the positive frequencies and discard the negative ones."[12]

Quantum field theory treats particles as waves travelling through fields that exist everywhere at once. Physicists use an object known as a *wave function* to describe the odds that a particle will be measured at any particular location or state of being. A particle's wave function can get complicated, with multiple waves interfering with one another to create patterns that do more to obscure information than reveal it.

Traditionally, scientists use a mathematical tool called a *Fourier transform* to deconstruct a complicated wave pattern into simpler constituent parts, allowing them to understand the relationship between, say, a particle's position and its momentum. Fourier transforms were well suited to Roger's sensibilities: they drew heavily on complex numbers and could best be understood geometrically.

Roger wasn't satisfied with Fourier transforms. They weren't simple or elegant enough for him. They involved too many equations. He could already see a simpler alternative: the Riemann sphere. Mapping the wave equations of quantum field theory onto the complex coordinate system of the Riemann sphere reduced the calculations required to separate amplitudes and turned an algebraic challenge into simple, beautiful geometry.

Though the technique was born from his stubbornness and talent, Roger was self-deprecating about it. He insisted conventional mathematics was just too difficult for him, forcing him to find an easier method. His modesty masked a growing confidence. He wasn't sure his method was completely original, but Engelbert and Ivor had never considered it. He had seen something nobody else in the kibbutz had.

He was also struck by how the Riemann sphere lent itself so well to both relativity and quantum mechanics. He returned to the idea that complex numbers might reveal underlying connections to unite the two. "They were clues to a deeper theory of how you might describe the world," he said. "When I learned about quantum mechanics, I was amazed by the fact that complex numbers are fundamental to the whole subject.... When I looked for complex numbers in General Relativity, you find them appearing amazingly and fortuitously in the equations for gravitational waves."[13]

He used his new approach to calculate the propagation of gravitational radiation in four-dimensional space-time. With simpler equations, the work progressed quickly, but he couldn't find all the solutions on his own. Ivor seemed like the best person to turn to. "I'd worked out some of the propagation equations, which were the most important parts. I was hoping somebody would collaborate with me to work out the whole lot. I pinned up the ones I had worked out over the door of Ivor's office, hoping he would pick up on them. He didn't pay any attention. It was Ted Newman who picked up on them and worked them out."[14]

Ezra T. Newman ("Ted" to his friends) was the son of a Brooklyn dentist. Loud and gruff, Ted always ready to volunteer his opinions on physics, politics, opera, and most other subjects. He laughed easily and uproariously. He loved telling rambling stories and playing practical jokes. He was two years older than Roger, though the two looked further apart in age.

Temperamentally, Roger and Ted made an odd pair. Each had a slight unease with the other's area of talent: Ted was most comfortable among the equations and formulas that so intimidated Roger. Conversely, he had none of Roger's geometric or visual skills. Nevertheless, Ted became Roger's closest friend and his most productive collaborator. "I have no idea why we became friends. Obviously, we had something. We had serious agreements on so many things from politics to music to easygoingness to judgements on physics," Ted said.[15]

They also bonded over their common dissatisfaction with the direction of current physics. "Before I had tenure, I used to tell people that I don't understand quantum mechanics," Ted said. "Faculty members told me, 'Shh! Don't tell anybody that. You'll never get tenure.' And then I met Roger and I whispered to him, 'I don't understand quantum mechanics.' And he whispered back, 'I don't understand it either.'"

Their professed confusion over quantum mechanics revealed as much ego as insecurity—they knew the problem wasn't with their capacity to grasp difficult concepts but with the theory itself. Roger truly did not understand why his colleagues seemed to treat their quantum theories as a complete and consistent scientific framework. Nobody had yet satisfactorily explained how quantum systems, with their superpositions and probabilistic measurements, worked. Roger knew he didn't know what actually happened on a subatomic level— and he suspected nobody else knew either. He found it reassuring to meet another physicist who was willing to name this ignorance.

"We resonated with each other because we had similar views about the world. I would say we had a similar view of scepticism. We were both quite prepared to believe that large bodies of people were going in the wrong direction, and that could be either to do with politics or physics," Roger said.[16]

Quantum theory was endlessly frustrating. The probabilistic nature of the quantum measurements implied that a particle might be in multiple places (or states) at once and only settle down into one location when an observer checked on it. Or the universe might split apart every time the wave function collapses, creating a separate reality for each possible outcome—the *many-worlds interpretation*.

Roger and Ted agreed the many-worlds interpretation was particularly preposterous. "The many-worlds people just don't have any clear connection between the theory they're proposing and the world we see around us," Roger said.[17]

Other quantum theories were just as counterintuitive. And it didn't help that the whole framework directly conflicted with relativity. Quantum theory described a universe composed of discrete, granulated quanta. General relativity treated space-time as a smooth continuum, sculpted by gravity. Much of the physics world was interested in reconciling the two by "quantizing gravity." Roger and Ted, both devoted relativists, believed the true solution would come from "gravitizing quanta."

They found more and more common ground. Like Roger, Ted had cared for and helped raise his younger sister, filling in some of the gaps of an absent father. Like Roger's, Ted's parents had expected him to follow his father's career, but love of physics had lured him away. And while they differed in approach, they both felt a near-mystical, aesthetic appreciation for the laws of nature they were trying to understand.

Ted loved natural beauty and took frequent hiking holidays in Utah and Alaska—an American version of Roger's walks in the woods. Both men were lifelong atheists, whose closest facsimile of spirituality came via the splendour of mature trees, dramatic geologic formations, and the equations of physics.

Roger, Joan, Ted, and Ted's wife, Sally, became a regular foursome. They camped together in Utah and Oregon, staying up late by the campfire talking about spinor networks and light cones beneath the starry sky. When they looked up, bathed in the ancient light of stars and galaxies that had blinked out of existence billions of years earlier, they saw more than most people.

Gravity created the stars and galaxies as well as the spaces between them. Nothingness had a geometry as real as the rocky outcrops towering over them. The atoms forming the stones and trees, the cloudy arc of the Milky Way, and the space-time curvature we experience as gravity (which could kill them if they took a wrong step) all emerged from the same laws. Maybe the shape of things created those laws. Maybe the laws created the shapes. Either way, the physical universe was beautiful on every level.

Even on a still night, everything was in motion. Planets and stars spun. Dust clouds collapsed, while some unknown force pushed each galaxy further away from every other. Nothing stayed the same for long.

Which made Ted and Roger's next major discovery even more mysterious. As they played with gravitational equations, certain numbers didn't seem to change no matter how they adjusted their parameters. For mathematical physicists, finding a constant was like discovering a new type of matter or a never-before-seen species of animal. A number that stays the same in a constantly shifting universe hints at something important, something fundamental. The speed of light. The force of gravity. The charge of an electron. The mass of a neutrino. Stars explode, firewood burns, rocks erode, empires fall, and marriages end, but these numbers stay the same.

Ted and Roger found ten new constants. They had no clue what these numbers were telling them. Some physicists said the Newman-Penrose constants weren't actually constants, and if they were, they didn't mean anything at all. Roger and Ted, though, were convinced they were on the trail of something important.

When Roger and Joan returned to England in 1961, he and Ted carried on their work and friendship through telephone calls, letters, and visits.

Complex geometry, conformal geometry, spinors, tensors, Riemann spheres, and the Newman-Penrose constants were taking him closer to the eternal nature of the universe. He was certain of it. Even more importantly, he was no longer exploring solo. He no longer felt so alone—not in this world or in the world-behind-the-world where new discoveries beckoned.

That same year, M. C. Escher took up Lionel's exhortation to incorporate the Penrose triangle and the Penrose staircase into his lithography. "Waterfall" riffs on the impossible triangle. It depicts a cluster of buildings on a hilltop connected by a viaduct. Water spills

M. C. Escher's "Waterfall."

over a central waterwheel, flowing back over two impossible tribars. The water appears to move downhill but ends up back at the top.

Escher's addition of detail and gravity turned Roger's spare visual paradox into a perpetual motion machine. "Waterfall" became one of Escher's most popular and iconic lithographs.

Roger felt more connected to the Dutch artist than ever. Examining a print of "Waterfall," he noticed the geometric ornaments on top of the viaduct's two towers. Each fused several regular polyhedra together, superimposing them at ninety- and forty-five-degree angles. Those spiky, multifaceted gems looked very familiar to Roger—he had built these very shapes out of cardboard as a child, sitting on the floor with his father. The similarity was coincidental, but it fed Roger's conviction that geometry was its own kind of reality. He and Escher—and Ted, Engelbert, Ivor, and others—were explorers, not inventors.

6

JABLONNA
AND BAARN

Roger and Ted Newman coauthored a paper titled "An Approach to Gravitational Radiation by a Method of Spin Coefficients."

Ted's influence was apparent: rather than deploying his typical graceful diagrams and pictures, Roger followed Ted's lead, filling the pages of the *Journal of Mathematical Physics* with the equations and formulas of tetrad calculus. Ted came incredibly close to a reputation-making discovery. But in one equation out of dozens in this highly cited paper, he erroneously wrote a plus sign instead of a minus.

"Unfortunately, a sign error in the equations prevented Ted from being the first to discover a beautiful solution of Einstein's vacuum equations now known to describe a rotating black hole," Roger wrote in his collected works.[1] New Zealand physicist Roy Kerr subsequently got the signs correct, and the result became known as the Kerr solution rather than the Newman solution. Ted didn't mind so much, but Roger never fully forgave himself for failing to catch the error.

• • •

Lionel and Margaret helped pay for a new home for Roger and Joan in Stanmore, a North London suburb at the very tip of the Northern Line. The modest home stood back from a cul-de-sac, with bushy trees spilling out from a wild, suburban forest covering the slope below. With neither Joan nor Roger at all prone to tidiness, every part of the house quickly became cluttered and messy.

A drive curved around the side of the house and down a steep slope to a one-car garage accessible only from the outside. Roger and Joan didn't own a car. The bare wood floor of the living room sat directly above the garage. Roger sawed a square hole in the middle of the floor, through a layer of asbestos insulation, into the garage below. He installed a trapdoor and ladder and furnished the cosy, dark space with a chair, desk, and lamp. He added a small notebook and a large blackboard—completing his toolkit for exploring the infinite depths of space and time.

He secured a new fellowship at London's King's College, studying the geometry of space-time and its relationship to Albert Einstein's theory of relativity. Among his fellow relativists, he continued to gravitate most readily to his Jewish colleagues, striking up a fast friendship with his officemate, a research fellow named Wolfgang Rindler. Seven years older than Roger, Wolfgang had fled Austria in 1938, arriving in England a few months before the outbreak of World War II.[2]

Wolfgang radiated gentle warmth and seemed motivated by pure love of researching, teaching, and communicating theoretical physics. Not having any of Roger's social awkwardness or seeming to grapple with self-worth, he had an understated self-assuredness that Roger admired. He embodied the unadulterated joy of working to understand the universe, without the self-doubt and familial complications that ate away at Roger. He had recently published a short book called *Special Relativity*, notable for its clarity and liveliness.

Roger's efforts to apply spinors to relativity captivated Wolfgang. "I'd heard of him, but never met him before. We talked a lot," Wolfgang recalled in an unpublished 2017 interview. "Roger was just in the process of inventing his spinor theory. He gave a series of lectures that winter. I attended those lectures and was fascinated by it. I took good notes of what he said. I found it incredibly beautiful. I'd never heard of it before and never seen it before. He sort of invented it as he went along."

Roger was still unused to having such close, uncomplicated friendships. He looked forward to getting to work each day, immersing himself in complexity and enjoying being somewhere he belonged.

On a spring day in 1962, Wolfgang turned to him and said, "Roger, you know this is terrific stuff. You should really publish a book."

"Well, I don't know how to write a book. I don't know how to begin it; I don't know the style. But you've written a book. Why don't you write it up?"

"Okay, let's publish it together."

They fell into a routine.

Roger appeared in their shared office carrying a stack of sheets covered with his formulas, diagrams, and theorems. The ideas would all be there but not in a presentable form.

"Now you Rindlerize this," he said.

Wolfgang organized Roger's thoughts into pedagogically useful order, leading readers step by step through proofs and insights. *Rindlerization* became their term for Wolfgang's transforming Roger's brilliant chaos and leaps of intuition into something readable and useful to others.

"Roger Penrose is a master in skipping all the mathematics. I had to do all the stuff in between," Wolfgang recalled.[3] "Roger has this incredible intuition. I have a little bit of that in physics but not in mathematics. It's a very different kind of intuition. It always bothered me that I didn't have it in mathematics. He had the intuition of what has to be true. I could prove it, but I couldn't intuit it."

They had nearly completed the first draft of *Spinors and Space-Time* before the end of that year. Wolfgang moved to Hamburg to work with another group of relativists. He took the manuscript with him, typing up final edits and mimeographing a few dozen copies.

Roger continued to thrive at King's, but his home life was much less satisfying. He began to dread the return to Stanmore, with its dreary clutter, small rooms, and spare furnishings. Joan couldn't seem to find friends or purpose. Loss seemed to have altered her permanently. The vicarious enjoyment she experienced with the Syracuse kibbutz didn't carry over to Roger's current colleagues. "The contrast between the mathematics scene in New York with its wonderful warmth and over here was completely different," she said. "I still look back very warmly at the times when we were in America."[4]

Roger demonstrated neither capacity nor willingness to understand Joan's unhappiness. He was on the brink of publishing a book and was respected by colleagues. He had friends and a future. He wished in frustration that his happiness would transform her, but Joan was perennially on the outside his success. And stuck in an unhappy marriage in early 1960s England, she could not find her own.

He chose a walking route from the Tube to the house through Stanmore woods, where he could pause and meander as much as possible. Since his childhood adventures in the Jungle, forests had freed his imagination and creativity. The twisted old trees and damp earth distracted him from the worries and responsibilities of domestic life.

At home, he'd disappear as quickly as he could, escaping from Joan through the living room portal, down the ladder into the garage, pulling the door shut behind him.

With a pencil or a piece of chalk in his hand, the low ceiling and dimly lit walls disappeared. His drawings extended off the page or the board, out of the house and beyond London's city limits, past the moon and the sun, Alpha Centauri and Andromeda, and on to the infinitely distant regions of time and space. If the woods freed his mind to explore aimlessly, the garage was where he got down to work, where his meandering curiosity became focused and precise.

The Syracuse group had gone their separate ways, but friendships and collaborations continued. Commercial air travel was becoming more common, making it easier to gather scattered researchers for international meetings, symposia, and conferences. Geopolitics made it more difficult for Soviet and Western scientists to work together, but even these barriers were starting to erode.

Andrzej Trautman had gone back to Poland to join the Institute of Theoretical Physics of Warsaw. He organized a conference that would

From left to right: Alfred Schild, an unidentified attendee, Jürgen Ehlers, David Finkelstein, J. Fletcher, Roger Penrose (standing on the running board of the car he and Joan drove to Jablonna), Dennis Sciama, Roy Kerr (almost hidden), Joshua Goldberg, and Rainer Sachs. From George Ellis and Roger Penrose, Tribute to Dennis Sciama for the Royal Society, Royal Society, 2010.

reconvene many kibbutz regulars, along with dozens of other physicists from around the world, to continue work on general relativity and gravitational radiation.

In summer 1962, Roger and Joan travelled by boat and train from London to West Berlin. After a day of sightseeing, they bought a Volkswagen Beetle at a suburban dealership. Nerves jangling, they drove up to a small wooden shack lined with sandbags: Checkpoint Charlie, a major crossing point between East and West Berlin. They passed through without incident and began a nearly 600-kilometer drive to Warsaw, Poland.

Their destination was the Staszica Palace in a Warsaw suburb where Roger would attend a monumental meeting known as the International Jablonna Conference on General Relativity and Gravitation. The event convened 114 of the greatest physicists from countries on both sides of the Iron Curtain, including 34 from the United States and Canada, 13 from the United Kingdom, 10 from the Soviet Union, 10 from East and West Germany, and others from Austria, Australia, Belgium, Bulgaria, Denmark, France, Hungary, Ireland, Iceland,

Israel, Italy, Romania, Switzerland, Sweden, Tunisia, and Poland.[5] It was the first time a group like this had met since before World War II.

"It wasn't just the science," Roger said. "East and West were getting together."

After many unsure steps—dropping biology, enrolling in maths, shifting to physics—he felt tentative joining this illustrious international community of physicists. Many of his friends were in attendance, which made it easier.

Ivor and Ted were there, as were Paul Dirac and John Wheeler. Others, like Leopold Infeld and Roy Kerr, Roger knew by reputation only. In Jablonna, he was also introduced to Peter Higgs, Subrahmanyan Chandrasekhar, and Vitali Ginzburg, all of whom were future Nobel Prize winners.

Towering above them all loomed Richard Feynman. Feynman was three years away from winning the Nobel Prize for his work in quantum electrodynamics and already had a formidable reputation as a researcher and celebrity scientist. He was as famous for his humour, tough talk, and womanizing as for physics. He was a legend among the men at Jablonna.

Roger had none of Feynman's bravado but connected with his skill at making physics visual. *Feynman diagrams* turned the arcane mathematics of interactions between subatomic particles into simple line drawings. With just a few squiggles and arrows, Feynman could tell the story of many finicky, hard-to-understand phenomena in quantum field theory. Subatomic particles weren't tiny marbles flying around and crashing into one another but excited states within quantum fields. Certain states in one field caused perturbations in others, and out of all these invisible, complicated interactions emerged the physical universe.

Physics had become far removed from what a human being could personally see and feel. Feynman diagrams made quantum field theory relatable and understandable—not just for laypeople but also (and especially) for scientists.

As an artist, mathematician, and physicist, Roger felt the power of Feynman diagrams. When the universe got too complicated, depicting it in a different way could reveal its unexpected simplicity. He had used two-component spinors to simplify the daunting, messy calculations of general relativity. Perhaps more visual tools like Feynman diagrams were waiting to be discovered and used to make the universe more comprehensible.

Simplicity was part of what Roger came to Jablonna to speak about. He and Ted gave a talk on *Newman-Penrose formalism*, their jointly devised notation system that used complex numbers to simplify aspects of relativity and required fewer, simpler equations to solve theorems related to gravitation, electromagnetism, and other areas. "The Einstein equations are mercilessly complicated," Ted recalled.[6] "With the Newman-Penrose formalism...you could solve them one equation after the other after the other. You could see what you were doing."

Like Feynman diagrams, their formalism was just a different way of presenting information, but it accelerated science and pushed it in more directions.

Roger also spoke about the *conformal compactification of space-time*, a mathematical technique he had developed to bring infinitely distant points into a finite reference frame, making it more possible to study the causal structure of space-time.

He shared one more blockbuster idea at the conference. Before and during the event, physicists continued debating whether gravitational radiation existed. Some attendees shared designs for experiments and instruments to detect gravitational radiation. Theoreticians like Roger focused on creating models and making predictions about how such radiation would behave.

Gravity was unsubtle when causing avalanches or binding galaxies together, but gravitational radiation couldn't be perceived by human senses. It was up to theoreticians to determine as precisely as possible how gravitational waves would behave, so experimentalists could know what to look for and how.

In his notebooks, Roger had been sketching light cones like the ones Dennis Sciama had shown him years earlier. The regions inside the future and past cones were considered "time-like." Events inside the cone could occur in the same place at different times. Outside the cone, things were "space-like"—events could occur simultaneously but in different places. Physicists used vectors to map these relationships. Every physicist in Jablonna was familiar with the distinctions between time-like and space-like vectors. Gravitational radiation might be time-like, space-like, both, or neither.

Andrzej had given talks in Syracuse on gravitational waves. Roger had found them too overwhelming. "He gave a lecture on the asymptotics of gravitational radiation, and it was full of lots of complicated calculations. And I thought, 'Well look, I can't follow all this. It's not my sort of thing. It just didn't appeal to me and my more geometrical way of thinking."

Staring at light cones rather than exponents and variables, he paid attention to a third region that most others disregarded: the cone itself. Vectors sitting exactly on the cone described the paths (or *world lines*) of massless objects moving at the speed of light. Rather than time-like or space-like, they were light-like. The future and past cones stretched out endlessly toward light-like infinity (also known as *null infinity*).

"There's a subtlety about it, which I hadn't appreciated. When you go to null infinity, the rate at which things fall off is different from when you go out in space-like directions," Roger said. He created a symbol for the infinitely distant region of the light cone—a handwritten, curlicued letter *I* (for infinity), which he dubbed *script I* and immediately shortened to *scri*. "People were just looking at equations and how things behaved when certain things went to infinity without thinking geometrically. I liked thinking of *scri* as a place that you could actually go and visit if you wanted to."[7]

Like a returning explorer, Roger regaled his Jablonna colleagues with tales of his journeys to *scri*, where gravitational radiation appeared to behave exactly as physicists hoped it would, radiating energy away in easily calculable, quantifiable patterns.

As he had done with his PhD, Roger forged ahead with abandon, lecturing and debating at a level of mathematical subtlety and density that dazzled his colleagues. His exploration of *scri* led to new insights about the *Sachs peeling property*, the multiple interpretations of *Weyl curvature*, and the conformal compactification of *Minkowski space*. Few people in the world could have appreciated Roger's geometric treatment of null infinity. Almost all of them were in the room with him.

To Roger's surprise and intimidation, he caught the attention of Feynman. Richard pulled him aside at the conference for a private talk to hear more about the spinor approach to relativity. The two spoke quietly and at length, with Richard asking piercing questions and actively absorbing Roger's responses. Roger walked away buzzing.

Most of the conference took place at the Jablonna palace complex, which included a network of parks and residences dating back to the 1700s. The German army had partially destroyed the building in 1944. When it was rebuilt, it became the property of the Polish Academy of Sciences. There was very little to see or do in the surrounding suburbs, which encouraged participants to stay close and socialize with each other.

Between lectures, scientists wandered the grounds, argued, and drank. Richard Feynman and Ivor Robinson played hopscotch with Iwo Bialynicki-Birula, Stanley Deser, Charles Misner, and Stanley Mandelstam. Joseph Weber gave a talk on his design for a new type of gravitational wave detector. Dmitri Ivanenko peppered him with questions so he could take the concept back to Moscow to develop a major Russian gravitational research project.

Engelbert Schücking and Alfred Schild, a Turkish-born Austrian American physicist who developed the first atomic clocks, were putting together a new research group in Austin, amply funded by the founders of Texas Instruments.[8] They approached Roger partway through the conference and offered to double his salary if he agreed to spend a year in Texas at their new Center for Advanced Study. They called it the "Princeton of Texas."

"In the summer of 1962, while attending Andrzej Trautman's relativity conference...we persuaded Roger Penrose, Roy Kerr, Ray

Sachs, Jürgen Ehlers, Luis Bel, and others to flock to the newly created Center of Gravity in Austin," Engelbert later recounted. This multi-national all-star team of relativity and quantum physics researchers would become the embodiment of American science. "Since Austin could not absorb all of Alfred's friends, he thought it would be good for Texas to have a second centre for relativity, and Alfred knew just the man to head one. When Lloyd Berkner, the boss of the newly founded Southwest Center for Advanced Studies, was looking for scientists, Schild was happy to advise him: Get Robinson."

With Ivor headed to Dallas, Engelbert in Austin, and so many of his other friends and colleagues also bound for Texas, Roger got swept up in the idea of going back to the United States. He finished the conference with a warm reception from his talks, a lucrative offer to join the University of Texas, and a sense of belonging that went beyond professional recognition.

Many of the participants planned to carry on from Jablonna to the resort town of Zakopone in southern Poland's Tatra Mountains. Roger and Joan, the only Westerners who came by automobile rather than government-sanctioned bus, wanted to join them, but the conference organizers refused to divulge the name of the hotel in Zakopone. Against the backdrop of the Cold War, they found this secrecy disconcerting. The Western scientists had been speculating throughout the conference about which Russian attendees might be spies. What international intrigue might they become embroiled in should they venture out on their own? It was unnerving.

"I mentioned this to Feynman about the hotel the day before we were to go, and he told us, 'I'll find out for you.' We were just about to leave, and he came up and handed Joan a little slip of paper. It said on the outside 'Agent X9' and inside was the name of the hotel. It was quite curious. I have no idea how Feynman found out," Roger recalled.[9]

Roger and Joan nervously pulled up at the place Feynman had sent them to in Zakopone. Very quickly, the mystery was revealed: the hotel was charging the physicists a group fee double the standard room rate. They didn't want independent travellers like Roger and Joan to secure

M. C. Escher's "Fishes and Scales."

their rooms at half the price—which they did, thanks to Feynman's surreptitious help.

Even more satisfyingly, Roger came away from the conference with his very own Richard Feynman story: their one-on-one conversation and the hotel adventure were a badge of honour and an indicator of Roger's bona fides in this rarified community of scientists.

Joan was fully invested in Roger's reputation. She told anyone who would listen what a great scientist he was. When the offer came to join the centre in Austin, she treated it as a given that they would go. Their lives were in the service of his career.

Before they left for Texas, Roger and Joan returned to continental Europe once more for a vacation. Joan was also pregnant again, which meant this might be their last solo trip together. They travelled to the Netherlands, renting a car and driving strategically close to

113

M. C. Escher's studio in Baarn. Roger rang him up from a public phone a few miles from his home and invited himself and Joan over.

Escher's two-story home was surrounded by trees, with a studio extension built out at the back of the house. In contrast to Roger's paper-and-chalk-dust-covered offices, Escher's studio was tidy and airy. He prepared tea for his unexpected visitors. Roger and Escher sat at opposite ends of a long table in the centre of the studio. A large window provided an expansive view of the surrounding forest. In the middle of the bench were two stacks of Escher's prints.

Using Lionel's jigsaw to cut out pieces, Roger had designed a wooden tiling puzzle to present as a gift. The puzzle had twelve identical pieces, each comprising a cluster of hexagons and partial hexagons that loosely resembled a tractor. The pieces could be rotated, flipped, and snugged together to create a larger, odd-shaped tile—like a hexagon with many bumps and other irregularities along its edges. Despite its odd shape, this larger tile could "tesselate" the plane—it fit together with copies of itself in a repeating pattern that stretched out forever.

Roger's puzzle had a specific geometric property that had not appeared in any of Escher's prints. It was *anisohedral*: even though the tiles were all the same shape and fit together in a repeating pattern, certain tiles had a unique relationship to the overall pattern and could not be flipped, rotated, or translated to match other tiles. It was the sort of geometric property only a very niche group of mathematicians generally appreciated. Escher did not consider himself a mathematician, but he was as fascinated with tiling as Roger. He later incorporated anisohedral tiling into the final painted work of his career, "Ghosts."

Escher gestured toward the smaller of two tidy stacks of recent prints on his workbench. "I don't have many of these left," he explained. He pushed the other pile toward Roger and said, "Choose one."[10]

Roger was overwhelmed and lingered over the decision of which to take as a gift. Eventually, he chose "Fish and Scales," a work also involving tiling but employing a different type of geometry than Roger's tractors. A large scowling fish glared at the right-hand side of the image. Each of its scales was slightly larger and more warped than the one to its right. As the scales grew in size, they morphed into more fish, growing larger and larger across the page, eventually becoming a second large fish glowering in the other direction. The scales of this fish curved and grew back in the other direction, growing into the original fish.

"Fish and Scales" used conformal geometry, that old favourite of Roger's that allowed shapes to infinitely scale up and down. Absolute distances and sizes don't matter in conformal geometry. The big and small fish in the image were conformally invariant. Viewers could continually change their frame of reference—"zoom out" their perception, resetting the cycle of expansion, and starting the process over again.

Escher seemed pleased with Roger's choice. "People don't normally appreciate that one," Escher told him.

When he got back to England, Roger framed the print and stored away new ideas about conformal geometry in the back of his mind. They joined his ever-growing repository of conceptual puzzle pieces—fragments of ideas he felt might lead him somewhere, either for work or play. Conformal geometry might be no more than a toy or it might contribute to new physics. He didn't care. It was fascinating either way.

Escher later noted his appreciation of Roger's unbounded enthusiasm for his work. "Young Dr. Roger Penrose, son of the London prof. paid me a very nice visit with his wife. We had so many things to discuss and so much to tell each other....I am often struck by the simplicity and childish playfulness of most of these learned scientists and that is why I like them and feel more at my ease with them than with my own colleagues," he recalled.[11]

Joan did not have the same optimism or excitement she'd had with her first pregnancy—becoming a mother would always be connected to that loss. Their son Christopher was born healthy in 1963, and the three of them prepared to take up the nomadic life that Joan had imagined two years earlier. Stanmore had been their home for less than two years, but they were ready for change. Roger would be a visiting associate professor at the University of Texas, back among his friends, who happened to form one of the most elite groups of theoretical physicists in the world. He hoped in Austin they could recapture some of the magic of Syracuse, which might at last lift Joan's spirits.

Joan's depression was a mystery to him, in a way that defied everything he knew about solving puzzles and enigmas. He found himself as incapable as Lionel of talking about emotional matters. Instead of trying, he disappeared through notebooks and blackboards into a place where he could work on puzzles he felt he actually had a chance of solving.

7

ASSASSINATION

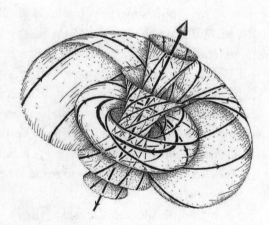

This Robinson congruence diagram depicts a family of intersecting light rays moving through four-dimensional complex space.

Austin was nothing like Princeton or Syracuse. Roger and Joan rented a detached, unfurnished home on a quiet street about fifteen minutes' walk from "the Forty Acres," as the campus was known. Every day shocked their sensibilities: wide streets, searing heat, huge cars, and guns seemingly everywhere.

"I absolutely hated Texas," Joan said. "It was the temperature to start, but it was really the violence. I remember there was a guy sitting in the main library. He had these Bermuda shorts or whatever you call them. He had his great big hairy leg on the table, and he had a revolver in his holster. When we went to buy furniture for the house, there was your bed, your dressing table, your mirror...and your gun rack. It was all knotty pine and everything. And that was Texas."[1]

Roger committed to Austin for only one year. He couldn't forsake London's twisted narrow streets, the comforts of familiar food and architecture, and the dusty old offices where his mind felt most free.[2]

Still, he revelled in reconnecting with colleagues from England, Syracuse, Princeton, and Jablonna. Engelbert Schücking had the office beside his. Roy Kerr and Ray Sachs worked down the hall in the other direction.[3] Three hours away in Dallas, Ivor Robinson and Alfred Schild led a second team that included Wolfgang Rindler and a Hungarian mathematical physicist named Istvan Ozsváth.

Wolfgang and Istvan were mild and quiet. Alfred's temperament was closer to Richard Feynman's. "Alfred was an ingenious mathematician who loved relativity and women. When interviewed by the Daily Texan, Austin's student newspaper, on how the professor conducted his research into the nature of the universe, he explained, frank as ever, 'I sit at my desk and think of girls—and sometimes I get a good idea,'" Engelbert later recalled.[4]

This kind of machismo permeated the Texas physics community. Roger and Joan could not rekindle the magic of the Syracuse kibbutz. Joan felt unwelcome and excluded by both the scientists and their wives. Overwhelming sadness sometimes layered down on top of the grief she still carried and the normal exhaustion of new motherhood, and she just couldn't get out of bed. Keeping up with arguments over gravitational radiation and two-spinor notation was too much.

Roger adamantly refused to let her unhappiness distract him.

He was at his most creative during the dark and silence of the early morning hours, long after Joan had gone to bed and Christopher was asleep. He'd often retire well after midnight and stroll to the office at 11:00 a.m. or noon, mind still working.

He perpetually returned to complex numbers and geometry.

Ivor had developed a new group of solutions to Albert Einstein's equations involving intersecting light rays moving through complex space-time that set Roger's mind racing with new ideas about the shape of the universe at its largest and smallest scales. "There was a business of finding special solutions to Einstein's equations. Ivor had a real talent for this. He was particularly interested in the 'null case,' in which you move one light ray into complex space. The intersecting rays twist around it, but not in complex space. They remain real. This was the kind of thing Ivor was good at doing—finding real solutions that seem to involve an imaginary light ray. I found these fascinating," Roger said.[5]

It was the kind of idea that would never make it into the popular press. The mathematics was simply too esoteric, the conclusions too far removed from any literal or metaphorical aspect of everyday life. People had enough trouble wrapping their heads around four-dimensional space-time without adding imaginary number coordinates into the mix. Only Roger, Ivor, and a handful of other mathematical physicists could appreciate how the poetic beauty of complex numbers might offer new predictions, reveal new knowledge, and raise new questions about gravity, radiation, space, and time.

Roger wasn't sure whether Ivor's solutions *described* reality or *were* reality. Complex space felt like a physical place he could visit. In fact, he liked going there. It had no restless baby, no grieving spouse, no guns, no ugly concrete-and-glass campus buildings. He could step out of Austin's harsh world to watch time stop at each Robinson congruence of light rays. He could follow their twists and spirals through real and imaginary spaces. He could perch on the spherical region of null infinity and try to pinpoint where reality emerged from it all.

If he could generalize Ivor's special solutions, he might be able to discover where space and time come from. The points where light rays intersected might themselves be the fundamental building blocks. Complex geometry could show how all these points fit together and build up into the classical and quantum worlds.

"There was some kind of complex world hiding behind the real world."[6] He was certain of it. "It always struck me as just an incredible piece of magic. You just put in that square root of minus one and a whole new world opened up. It wasn't a theory of the world yet. But I thought of it as a theory of the world."[7]

This quest kept him awake at night, occupied his mind on the walk to work, and was the first thing he addressed each day when he arrived. It teased and frustrated him. The breakthrough he sought always seemed to be just out of reach, hidden behind a door he had not yet found a way to open.

On the morning of Friday, November 22, 1963, Roger walked to his corner office in the Graduate Research Center. He arrived late, as usual, having stayed up too late, as usual. Nobody minded. The men who worked at the center were paid to be brilliant, not prompt.

Out his window he could see "the Tower," a twenty-seven-storey admin and library building that dominated the campus.[8] The physicists were temporarily housed on a floor of the university's new Business Economics Building (BEB) (students immediately dubbed it the Big Enormous Building).[9] Opened in March of that year, the massive BEB was an architectural showpiece, with many special touches designed to position the campus at the forefront of "push button society." The facade featured fifty multicoloured button-motif panels that echoed the advanced computing and vending machine technologies available within. The building also featured the university's first bank of escalators, endlessly moving staircases transporting people up at high speed to classrooms and offices. (Unlike Escher's print, the BEB escalator system only ascended. To get back down you had to take old-fashioned stairs.)

Roger preferred the stairs in both directions. He tried to walk as much as possible to keep his weight down.[10] He instinctively disdained huge, modern university buildings. He called them "research factories" and felt they lacked the warmth and intimacy that fostered creative work.[11]

He set his briefcase down next to disorderly piles of papers that obscured his desk. His chalkboard was densely covered with sketches and diagrams. Arcs ran from the bottom to the top, charting particles' world lines through space and time. Light cones extended into the past and future, tips touching at a single point—an arbitrary "now."

Light rays travelled along the sides of the cones, the null region, at the fastest speed physics allowed. Calling this limit—299,792,458 meters per second—*the speed of light* is correct but not quite accurate. More precisely, light in a vacuum moves at the *speed of causality*. Other phenomena, including gravitational waves, travel at the same speed. Exceeding the speed of causality would create paradoxes in which events happened before the circumstances that caused them.

Past light cones contained every possible cause that could have affected a given "now." Future light cones held every effect that "now" could possibly influence. But what was "now"? Did the present moment have unique physical properties that differentiated it from past and future? Did it have a shape or a size? How did all these "nows" fit together to make a universe?

Roger crowded complex equations around his diagrams, trying to turn his fascination into insight.

He wasn't convinced he was asking the right questions. Breaking the universe down into its basic units of matter and energy, gravity and electricity, space and time, cause and effect was meant to get at the very essence of reality. Could there be something even deeper? Even more fundamental? What if the universe wasn't made out of space-time, but space-time was made of something else?

A telephone started ringing in an empty office down the hall. Roger turned away from the board and then turned back, consumed with Ivor's congruences. Ivor was particularly on his mind that day because he and the rest of the Dallas group were distinguished guests at a lunchtime reception at the Dallas Trade Mart where US president John F. Kennedy was to speak.[12]

The race between the United States and the USSR to reach the moon had captured the public imagination, and theoretical physics was enjoying an upswell of financial and public support. Kennedy, who was preparing to announce his campaign for a second term, had come to Texas to talk about the future of American science. The event was organized by the Dallas Citizens Council and the Graduate Research Center of the Southwest, which funded Roger and all his colleagues. Kennedy was scheduled to travel to Austin after the reception for a fundraiser.[13] Joan had dressed nine-month-old Christopher in his best clothes and planned to take him on her shoulders to greet the motorcade.

Kennedy chose Texas to talk about science for a reason. The team Ivor, Engelbert, and Alfred had assembled was the envy of Princeton, Cambridge, and Moscow. It had very quickly become a source of national pride.

"There is no longer any doubt about the strength and skill of American science, American industry, American education, and the American free enterprise system," Kennedy's prepared remarks read. "In short, our nation's space effort represents a great gain in, and a great resource of our national strength—and both Texas and Texans are contributing greatly to this strength."

As his Dallas colleagues were finding their table at the Trade Mart, Roger was enjoying the silence of the still largely empty physics floor in Austin.

He was at peace, standing alone at his chalkboard. Alone at infinity. Alone in complex space. Toying with real and imaginary numbers, working with quantum and relativistic physics. Remembering his father's lesson that a puzzle was a puzzle whether it involved freeing a ring from a loop of string or answering the hardest questions in modern physics. The only real difference was that a toy puzzle was guaranteed to have a solution. Roger couldn't be completely certain that the solutions he sought even existed.

A phone rang in another office. It too went unanswered.

Reconciling physics' two major theories had become the field's single most important and elusive quest. Relativity did an excellent job of describing "macroscopic" objects like stars, people, and bullets.

Quantum physics provided an equally powerful set of laws dealing with the tiny world of subatomic particles and forces. But large objects are made of tiny ones, and the two theories described the same physical reality: Roger Penrose was both a 5'6" human being whose body obeyed the laws of classical physics and also a collection of about seven octillion atoms governed by the entirely incompatible laws of quantum mechanics.[14] One interpretation was no more real or fundamental than the other; each must reflect some as-yet-undiscovered deeper theory that absorbed them both.

Gravity was at the heart of the incompatibility. Relativity was a theory of gravitation. Quantum theory excluded it. If gravitational radiation existed, nobody knew what it looked like on a quantum scale.

Physicists had many ideas about how to bring quantum mechanics and relativity together, but Roger found most of them full of intolerable hand waving. He preferred dealing exclusively with reality, which somehow led him ever back to imaginary numbers.

These numbers felt as real and palpable as stars and stones. Even the square root of negative one was a tool he could hold in his hand like a piece of chalk.

A third phone rang. Another scientist had finally arrived and picked up.

"I saw Ray Sachs rushing into the corridor white as a sheet," Roger said. "Ray told me, and then I phoned Joan. She hadn't heard about it at that stage. It had just happened."

The noon hour Roger spent contemplating Robinson congruences coincided with the harrowing experiences of his Dallas colleagues.

President Kennedy had been scheduled to address more than 1,000 people at the Trade Mart, including local politicians and business leaders. The Graduate Center had reserved a table for Wolfgang and Phyllis Rindler, Istvan and Zsuzsanna Ozsváth, Ivor Robinson, and others from the Dallas relativity group.

"Amongst all these notables, there was the entire faculty from the centre," Wolfgang later recalled.[15] "It became twelve thirty when he should have been there, and he still wasn't there. There were some rumours in the hall that things would be delayed."

Outside the hall, panic was spreading. Inside, people merely grumbled.

"I didn't eat that morning and I was rather hungry," Zsuzsanna recalled. "I said, 'He could be more on time a little bit.' Then suddenly Erik Jonsson went to the podium and used a strange word: 'mishap.'"

J. Erik Jonsson was president of the Dallas Citizens Council, cosponsor of the event. "Ladies and gentlemen, may I have your attention, please?" Jonsson said from the podium at 1:01 p.m. "There has been a delay in the arrival of the motorcade. There has been a mishap. We do not know the extent of it or the exact nature. We believe from our report that we have just received that it is not serious. We hope you will keep your seats. As soon as we have something to tell you, believe me, we'll do it."[16]

The physicists and their spouses assumed car trouble or some other inconvenience. The food workers had a radio in the kitchen. The serving staff were among the first people in the building to know what had happened.

In a moment of logistical absurdity, someone decided there was no longer any point in waiting for the president and gave the order to start lunch service. A server laid a plate of steak in front of Zsuzsanna.

"What happened? Why do we get lunch now?" she asked.

The waiter pointed a finger at his head. "Bang, bang." He went back to the kitchen.

It didn't seem possible that someone could deliver such devastating news alongside a T-bone. Zsuzsanna thought it was a disgusting attempt at humour.

"I am not sure he made jokes," Wolfgang said.

"What do you mean?" she asked.

"I am not sure he made jokes."[17]

Plates were still emerging from the kitchen when Jonsson returned to the podium twelve minutes later.

"I'm not sure that I can say what I have to say," he said. "I feel a little bit like the fellow on Pearl Harbor Day. It is true that our president and Governor Connally in the motorcade have been shot."

Confusion and disbelief spread through the room.

"In those days people didn't have cell phones and so once you were sitting at lunch, you had no connection with the outside," Wolfgang recalled. "It was ridiculous. People were eating there, and Kennedy was already dead."[18]

Wolfgang had fled Austria in 1938, the year it was annexed by Nazi Germany. The Ozsváths' view of the world was shaped by political violence in their native Hungary.

"Me and my husband grew up on the murder of the future king in Hungary in 1914"—an assassination that precipitated World War I—Zsuzsanna said. The moment she heard Kennedy had been assassinated, America stopped feeling like a haven from such cataclysm. "My only thought was there was another war coming. I was terrified that we were going to be in a war."

Back in Austin, Roger left his chalk, his numbers, and his stacks of paper.

Physically, this "now" was similar to any other. It had past and future light cones. Over billions of years, events had tumbled and swirled and interacted to result in this moment. The ripple effects of whatever came next were already spreading out in a spherical wave, expanding at the speed of causality.

Roger, though, had temporarily lost his ability to step out of this world into the neutral clockwork of cosmic cause and effect. This wasn't just another point where the tips of two light cones touched.

In a fog of reality and unreality, he walked out of his office, left campus, and returned home to find Joan. Kennedy's assassination temporarily swamped the tension between them. They stayed by the radio, listening for updates, calling friends, and grieving.

Two days later, Kennedy's assassin, Lee Harvey Oswald, was himself shot and killed by a nightclub owner named Jack Ruby. Like Zsuzsanna, Joan and Roger were terrified. Everything seemed to be spinning out of control. "It did seem as though it was the whole place just going to pieces and everybody was being shot," Roger recalled. "It really was an appalling time."[19]

Their grief mixed with disgust. Roger's antipathy for Texas grew unbearable. "I remember looking out the window, and seeing some young kid playing with a gun and making fun of the Kennedy assassination," he said. "I thought this was terrible. And it was an indicator of the attitude of Texans toward guns that that kid's parent wasn't horrified by that child's behaviour."

For once, he and Joan saw the world the same way. "There was at least one party on our road...celebrating the death of Kennedy," Joan recalled with distaste in a 2019 interview. "Texans were out to get him."

Austin felt hostile and alienating. They drove to Dallas, seeking the company of old friends. Roger and Joan, Zsuzsanna and Istvan, and Wolfgang and Phyllis decided to flee. The three couples drove in two cars across Texas to San Antonio and on to the Atlantic Ocean. In Corpus Christi, they spent nearly a week staring at the surf, processing Kennedy's death, and speculating about what might happen next.

Joan and the others were still reeling as Roger began to regain his equilibrium. Time still passed. Gravity still kept his feet on the ground. Stars still radiated light across the night sky. He instinctively withdrew from the intense emotional conversations among his travelling companions. He found himself desiring silence, evenness, and a return to contemplation.[20]

Ordinary life also intruded on Joan. The six friends were there to grieve these awful events together, but Joan sensed she didn't belong. Though she and Roger had only been married a few years, their relationship was visibly strained. Roger's friends and colleagues picked up on this tension and unkindly assumed Joan was the problem.

"Joan was a rather good looking, interesting woman, but not particularly highly educated, if I may say so," Zsuzsanna said.[21] "Everybody was dizzy from how great Roger was. Everybody talked about that constantly. Every man, woman, and child who was around spoke about 'the greatness of Roger' and she very much felt pushed into the back. I don't really think that it could be called a good marriage if I am honest."

In Corpus Christi, Roger just wanted to get away from Joan—and everyone else.

When it came time to leave, Istvan offered for Roger to ride back to Dallas with him, allowing the others to talk and gossip to their hearts' content. Roger welcomed the invitation. "He was a kind of silent chap. Perfectly nice, but didn't talk much. And so I had this long time of being able to think," Roger recalled.[22]

They rolled along Texas highways in comfortable, sombre silence. Roger's mind returned to light cones and twisting rays. He felt peace returning. He was back at null infinity, back in complex space, back where things made sense. The chaos of the preceding days fell away. He viewed the puzzle he had been working on for so long with fresh eyes.

"I decided that the time had come to count the number of dimensions in the Robinson congruences. I was surprised that the number of real dimensions was only six (so only three complex dimensions)," he later wrote. "I had found my space! The points...indeed had a very direct and satisfyingly relevant physical interpretation as 'rays', i.e. as the classical paths of massless particles. And the 'complexification' of these rays led, as I had decided that I required, to the adding merely of one extra real dimension."[23]

Riding along with Istvan silent beside him, Roger saw in a flash how to generalize Ivor's solutions. Out of that complex math, four-dimensional space-time (the very space-time in our universe) naturally emerged. In his new solutions, each space-time moment—where a family of light rays intersected or past and future light cones met—was a complex sphere rather than a zero-dimensional object. Light rays twisted over the surface of the sphere creating vivid, complex twists and spirals. Those spheres combined to form space and time as we know them.

Somewhere between San Antonio and Dallas, he understood how every particle, every force, every light cone, every cause, effect, law, and property might be, in its most fundamental form, pure, glorious complex geometry.

He didn't have the name yet, but he had just developed the foundations of *twistor theory*, a framework for reality that would occupy him for the next six decades and become the life accomplishment he most wished to be remembered for. He built complexity on top of complexity. Twistors were constructed from pairs of spinors. On-ramps and overpasses flew past as Roger imagined how this jumble of real and imaginary numbers transformed into a sensible structure, morphing from abstract mathematical relationships into space, time, matter, energy, and, ultimately, an entire universe.

The Robinson congruences were part of something vast: a new, stunning picture of reality. And at that moment, Roger was the only person in the world who understood it.

A door had opened in his mind. Beyond it, a path as wide and clear as a Texas highway lay before him. A twistorial reality could explain everything from the spin of massless particles to the nature of gravitational radiation. It could unify general relativity and quantum mechanics.[24] It could remove the final layers of illusion to reveal the universe's true nature. Given that many quantum theorists believed physics could not describe "real" objects but only our subjective experience of them, Roger's conviction was even more extraordinary.

He forgot the recent horrors that put him on this road trip. He forgot about Joan, travelling behind him on the same highway in a very different frame of mind. The tragedies of this world faded into the background, overwhelmed by his unexpected, serendipitous inspiration and renewed sense of purpose and possibility.

"I got back to the house in Austin and got to my blackboard. And I thought, 'How do I write these things down? Well, best thing I know about is two-component spinors. That's the way you do it of course. Of course you do.' So the light rays were defined by two-component spinors, and you have to push them into complex space. I wrote it down, and it just flopped out like that. It was twistors."

His discovery felt miraculous. He had been exploring these ideas for years, calculating, discussing, reading. The solution appeared when his mind was somewhere else entirely.

"On the trip back to Austin, in that long silence, I was able to think. That insight came from the good fortune of being in a car with someone who was not talkative. The moral of the story is that there are lots of ways to get inspiration," he said.[25]

The next morning, Roger once again arrived late, having once again stayed up too late.

He wiped his blackboard clean and began drawing anew, sketching out the ideas that had come to him the day before. His concentration was broken by a conversation coming from the office next door. Roger wandered over to find Roy Kerr and Ray Sachs engaged in an intense conversation in front of Roy's blackboard. Roy had found a type of complex analytic expression that generalized Ivor's special-case solutions for twisting light ray congruences in complex space.

Roger couldn't really grasp what Roy was saying without a visual explanation.

"What are these expressions? What do they *look* like?" he asked.

Roy swiftly chalked equations and diagrams onto the board that Roger immediately recognized.

"My God! Those are twistors!"

Even though he had arrived at a similar insight within days of Roger's own epiphany, Roy didn't share Roger's sense that twistors might be the key to every question in modern theoretical physics. In fact, very few of his colleagues shared his excitement.

"There were only about a dozen people in the world who could understand what I was trying to do with twistors, and very few of them understood why it mattered," Roger recalled.[26]

Roger had come far enough in his career, though, to trust his own instincts. He finished his year in Texas with a potent new direction in which to take his research. He was desperate to leave Austin, to return to England, and continue his work.

8

THE SKY IN
A DIAMOND

Joan and Roger took Christopher on one more trip to Blink Bonnie before they left North America, to escape the heat, drop off some of their possessions, and see if they could locate one more dose of peace and quiet before the next chapter began.

"Roger is very good at space, and I remember he packed our little Volkswagen Beetle so there was just enough room for Christopher in the back. It was a little Christopher-shaped hole. We drove all the way to Canada and then we drove all the way back, just to escape. That was that," Joan said.

They returned to North London in 1964. Roger found work as a reader—a step above senior lecturer and below professor—at Birkbeck College in the Bloomsbury neighbourhood.

He was relieved to be back in England and excited to be employed as a physicist. His Birkbeck office had an even larger blackboard. It was crammed with papers and journals, including stacks of marking and student evaluations that Roger was constantly behind on. Any time spent organizing or clearing space was time away from the chalkboard, a sacrifice he was rarely willing to make.

He felt at home at Birkbeck: productive, intellectually stimulated, appreciated. He liked his students and liked teaching—it seemed to help his own thinking to explain his ideas to a receptive audience.

Twenty-two months after Christopher was born, Roger and Joan had a second son, Toby. The house was messier, louder, and more chaotic. Without the distractions of living abroad, their daily routines became tense and grating. Joan was at a loss socially and professionally. She couldn't find a way to fit in with Roger's friends, colleagues, and family.

Roger also had pangs of sadness. He wondered what life would have been like had their daughter lived. Still, he liked being a father. He cooked for the boys, took them to the park, read them books like *The Phantom Tollbooth*, and, as best he could, shared his own passions with them. "Roger was teaching Christopher negative numbers when he was two. I don't know if Chris still remembers that, but I do," Joan said.

Joan's depression irritated Roger. Twistor theory was gnawing away at him, and it bothered him that she wasn't more present, more excited about his ideas, or more willing to free him up from parenting and housekeeping duties.

He had Lionel and Margaret's voices in his head, warning him not to marry Joan. He now realized the folly of ever thinking he could resolve Joan's depression through his own success and stability.

That was naive. Well-meaning but naive. I didn't have relations much with women. Since she seemed to be presenting herself as someone I could have the relationship with, I wanted to have a relationship. She was an interesting person, and she certainly had talents. She was an intelligent woman, but I was never really in love with her. It was a bad mistake to marry her. She had psychological problems, and I thought these were things I could at least help to sort out. Having a stable relationship with me, so I thought, would help them.[1]

The more he thought about it, the more he cleaved to the self-serving idea that Joan had duped him. He only knew about physics. Relationships were Joan's area of expertise—she liked saying she "majored in boys," after all. Surely she had taken calculated advantage of his naïveté and tricked him into this unhappy marriage.

Here they were with a house and two children; his work was going very well, and he had a bright future ahead of him. His life was better than stable—he was thriving. His papers on conformal infinity and the spinor approach to relativity appeared in prestigious physics journals. He received more and more invitations to give lectures and attend conferences. He could not fathom why Joan was still so unhappy.

They argued over household matters, finances, and social activities. They both left the house in perpetual disarray. Their social life was strained. Joan felt helplessly alone. Roger sought out solitude. He craved another hit of that magical connection he'd had on the silent ride through Texas.

Late at night, after his family had fallen asleep, Roger habitually slipped through the trapdoor, out of this world into another. The ceiling was low, the light dim, the air cool. The world had less of a hold on him down here. The house, the cul-de-sac, and the woods beyond were still. Silence settled on him. Distracting thoughts of dirty dishes and broken toys, his latest spat with Joan, and nagging insecurities about his career and marriage became muffled and then disappeared. He could breathe and think and imagine in unadulterated peace.

This was his world, not the noisy chaos above.

Lionel had once shown him a sundial, and in that moment, his life and everything he knew became tiny, dwarfed by the clockwork mechanics of the solar system. Now he had a universe to play with: endless space filled with stars and galaxies and who knew what else. His problems with Joan became a tiny dot, lost within this magnificent sweep of cosmological mechanics.

Roger shifted his attention to phenomena so large, small, distant, dense, fast, ancient, bright, and just plain strange that they were completely removed from the everyday realities of work and home. These visions touched the foundations upon which everything familiar rested. Each door Roger pushed open took him further from a life he was coming to detest. He escaped reality by venturing deeper into it.

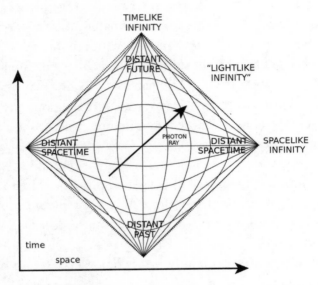

Penrose diagram.

On one such night, he switched on his lamp, releasing a stream of photons illuminating a blank notebook page. The electromagnetic forces that bound the atoms of his pencil made it feel solid and comforting in his hand.

He drew a square tilted at forty-five degrees, corners pointing toward the top, bottom, and sides. He connected opposing corners with straight lines, one vertical and one horizontal, intersecting at the diamond's centre. He felt himself outside and inside the diamond at the same time.

Drawing freehand, he added successive curved lines connecting opposite vertices, obeying the graceful rules of conformal geometry:

preserving angles where each arc intersected, shrinking the regions they circumscribed as they approached the edges. The grid seemed to bulge in the middle and bend away near the perimeter.

A few years earlier, M. C. Escher had experimented with similar grids, filling the infinitely scaling regions with lizards, angels, devils, and fish. Roger filled his with the entire contents of the universe.

Like light cones, his diagram depicted the four dimensions of space-time on a two-dimensional surface. The horizontal axis represented three spatial dimensions. Time ran up and down the page. The left and right points represented infinitely distant regions of space—space-like infinities. The top vertex was the infinite future and the bottom the infinite past—time-like infinities. As with a light cone, rays of light travelling across the universe would chart perfect forty-five-degree lines across the diagram—parallel to the edges of the diamond. The curves were like lines of latitude and longitude, a coordinate system to track the movement of light and causality between different points in space-time. Due to the diagram's conformal scaling, a light ray would take precisely forever to travel from the centre to the edge—approaching a light-like or null infinity.

His bulging diamond bounded an infinite four-dimensional universe within a finite 2-D shape. In those conformal curves, Roger could map the causal relationships between any two points in relativistic space-time. All the exploding stars and collapsing gas clouds, the crowded cities of Earth and the barren regions of intergalactic space, the silence of the moon and the noise of his own household, connected by one orderly drawing, where they could be charted and understood.

Roger put down his pencil.

This diagram was a marvel of simplicity. It offered a new way to talk about and understand relativity. The mathematicians and physicists who contested and debated relativity didn't always share terminology or even speak the same language. The Penrose diagram was an incredibly powerful visual tool that allowed physicists, "formalist" mathematicians, and geometers like Roger to talk to each other about relativity. Roger felt he'd tapped into something even deeper: perhaps he had discovered the true shape of the universe.

He hadn't, quite. His diagram depicted a universe filled only with massless light. Our universe is chock-a-block with matter, which doesn't create the same kind of conformal geometry. He drew the universe as infinitely old and infinitely large, neither of which were sure things. Still, Penrose diagrams had something powerful. He would have to explore in greater detail. This was just the beginning.

For now, he savoured the drawing in front of him. Like an ancient seafarer reaching an unexplored island, he gazed down at pristine territory. In the years to come, Penrose diagrams would be folded, torn, rotated through three dimensions, combined, and distorted to decode the universe's most extreme geometry. At this moment though, Roger was alone with his notion of untouched geometric purity.

He was both energized and at peace—the exact opposite of how he had entered his office. Work done for the day, he turned out the light, climbed back up into the ordinary silence of his sleeping household, and went to bed.

9

THE PENROSE
SINGULARITY
THEOREM

In autumn 1964, Roger Penrose crossed the street near his office at Birkbeck College in central London. As he stepped from one kerb to the other, he experienced a mental flicker—a momentary thought that disappeared as soon as he reached the far side.

That moment changed the course of cosmology for decades to follow.

Two and a half billion years earlier, in a remote region of space out beyond the stars of the constellation Virgo, a light source four trillion times brighter than our sun poured radiation out into space. A sphere of electromagnetic radiation blasted out in every direction at once at a speed of nearly 300,000 kilometres per second.

Our planet then looked nothing like it does now. Single-celled organisms were relatively new, and more complex creatures were still hundreds of millions of years away.

As that light travelled, life on Earth developed. Along came sponges, fungi, and other multicellular organisms. When the radiation was about 540 million light years away, a burst of speciation known as the Cambrian explosion produced most of Earth's modern animal categories. Fish filled the seas, and insects swarmed the skies. Continents drifted; dinosaurs came and went.

Much later, when the light was just 2.5 million light years away, *Homo habilis* began making and using tools. The radiation was only 300,000 light years away when *Homo sapiens* appeared.

Civilizations and empires rose, clashed, and fell. Ancient Sumerian and Egyptian societies developed written languages and systems of mathematics and logic. As humans explored the mountains and the oceans, they also turned their eyes skyward to study the motions of the sun, moon, and stars.

When London was still new, before Birkbeck College existed, when the light was still on its way, new scientific insights cascaded one after the next. In the sixteenth century, Nicolaus Copernicus proposed that Earth moved around the sun rather than the other way around.

In summer 1666, an apple fell in the Newton family orchard. The fruit lay still on the ground, but Isaac Newton's mind was in motion. Some force caused the apple to fall. Some force caused everything to fall. And what if that force also kept the moon circling Earth and Earth circling the sun?

In 1687, Newton published *Philosophiae naturalis principia mathematica*, three books of scientific and mathematical laws that included a theory of universal gravitation. Every object—be it planet or plantain, star or starfruit—exerted a pull on every other. Newton connected his family orchard to celestial mechanics, demonstrating that the ethereal cosmos could be understood through scientific investigation.

In 1676, Danish astronomer Ole Roemer measured the speed of light—putting to rest earlier theories that the movement of light might be instantaneous.

In the 1800s, James Clerk Maxwell theorized that there must exist a type of low-frequency, long-wavelength electromagnetic radiation that would later be called *radio waves*. A few years later, Heinrich Hertz produced, transmitted, and received radio waves for the first time.

Scientists built instruments to detect what the human eye could not. An invisible world opened up of radio waves, microwaves, infrared radiation, ultraviolet light, X-rays, and gamma rays. Science was approaching the point where it could detect that burst of energy, still journeying toward Earth.

In the early 1900s, German-born physicist Albert Einstein, working not in a laboratory or observatory but in the realm of thought experiments, imagined how the universe would look to a particle of light. He determined the speed of light was constant for all observers in all frames of reference.

In 1905, Einstein published his special theory of relativity, describing how time slows as matter approaches the speed of light. Mass could be converted into energy and vice versa. Space and time were not separate phenomena but a single space-time continuum.

Ten years later, he published his general theory of relativity, which treated gravity as a warping of space-time. The Earth, he said, holds the moon in orbit not by pulling on it but by bending space-time so much that the moon's path becomes an ellipse.

Out of Einstein's theories came light cones, space-time diagrams, Penrose diagrams, and a century of advances in physics.

Just over a month after Einstein published his theory, a German physicist name Karl Schwarzschild found a solution to his equations that described the curvature of space-time around a perfectly spherical body. Schwarzschild showed that relativity allowed for the existence of regions so dense that space-time would fold in on itself, becoming so warped not even light could escape. Physicists acknowledged Schwarzschild's math was correct, but few believed such a *singularity* could exist in reality.

Schwarzschild's model required hyper-idealized circumstances. The real world appeared too flawed and messy to produce a Schwarzschild singularity.

More papers followed. In 1939, J. Robert Oppenheimer and one of his students published a paper arguing that "when all thermonuclear sources of energy are exhausted a sufficiently heavy star will collapse.... This contraction will continue indefinitely." But their singularity calculations also concerned an "idealized case" of a perfectly spherical star with no internal pressure—not like any actual object.

The story always seemed the same: catastrophic gravitational collapse could only happen in circumstances too specific, too symmetrical, too idealized to have any bearing on the real world. The very idea that relativity, the most complete physics theory ever known, could predict circumstances under which it would itself break down seemed absurd.

In 1932, a Bell Telephone engineer named Karl Jansky became the first scientist to point a radio wave antenna at the sky, launching a new era of radio astronomy. In the late 1950s, radio astronomers detected a powerful source of radiation emanating from the direction of the constellation Virgo. That light, finally at the end of its epic 2.5-billion-year journey, with its dying gasp, created a blip in their data. That blip set off a massive debate and an international race to explain its origins.

The radiation source was unlike any star, galaxy, or other known object. It was brighter than a galaxy, with an unrecognizable pattern of wavelengths.

This exotic cosmic beacon, this source of bafflement and intrigue, this radiant celestial enigma, was prosaically catalogued as C3 273—one of several recently discovered radiation sources tentatively categorized as *quasi-stellar objects* or *quasars*.

California Institute of Technology astronomer Maarten Schmidt proposed that the strange pattern of radiation wavelengths might result from "red-shifting," which would mean the object—whatever it was—was likely moving away from Earth extremely quickly. This would mean it was very distant and very old. So much energy packed into such a tiny source suggested the impossible: this radio source might hide a singularity.

Further observations confirmed the redshift but settled no arguments.

Could quasars be larger and closer than they seemed? Could the red-shifting result from something other than it speeding away? And if they really were so small and distant, where was all that power coming from?

Looking through historical records, astronomers discovered that quasars had been showing up unnoticed on visible-spectrum images for decades. These archival images indicated quasars grew brighter and dimmer from one year to the next—shockingly fast changes for a cosmological object. More data revealed these brightness shifts could happen in a matter of hours. A galaxy-sized object can't have a dimmer switch that works that quickly. Because nothing can move faster than the speed of light, it is physically impossible for an object millions of light years across to grow brighter and dimmer that quickly. Whatever quasars were, they appeared to be only a few light hours in diameter.

As uncomfortable as it was, quasars fit every description of a far-distant singularity in the final stages of devouring a galaxy's worth of matter and energy. No light would escape from the singularity itself, but the surrounding region would be incandescent.

Could singularities be real after all?

All around the world, physicists scrambled to prove either that singularities were more possible than they thought or that some other relativity-compatible explanation could account for quasars.

In Austin, Princeton, and Moscow, at Cambridge and Oxford, in South Africa, New Zealand, India, and elsewhere, cosmologists, astronomers, and mathematicians ran scenarios, changed variables, held conferences, and floated ideas, trying to create a definitive theory of quasars.

Most scientists looked for specific solutions to Einstein's equations, trying to identify the exact circumstances that could create objects like C3 273. They wanted an answer as specific as Schwarzschild's or Oppenheimer's. Roger believed searching for special solutions was a dead end.

"That was the sort of trickery and mathematical ingenuity that people used to find solutions. They were not generic solutions. If you have a general space that is collapsing in a complicated way, you've got no hope in hell of finding exact solutions that would capture that," Roger said.[1]

The quasar conundrum sparked Roger's geometric creativity and his instinct for unexpected simplicity. Since his days making perpetual calendars and moon clocks, he had gravitated toward underlying principles and universal structures. He didn't want to find some outlandish set of circumstances under which a singularity could form. He wanted a general rule with no need for idealized circumstances.

He could fold, rotate, bend, and stretch shapes in his mind, imagining the results visually, rather than as equations. As he worked on the problem, he realized he had a more specific advantage as well: he was onto an idea few other theoretical physicists were thinking about.

A few months earlier, he had written a paper that was still under review with the *Proceedings of the Royal Society of London*. The paper wasn't directly related to singularities or event horizons. Titled "Zero Rest-Mass Fields Including Gravitation: Asymptotic Behaviour," it was a detailed description of how he could use his beloved conformal geometry to characterize the general radiation properties of space-time points within gravitational, electromagnetic, and neutrino fields. An appendix to the paper introduced a concept Roger would come to call a *closed trapped surface*.

On most curved surfaces, light rays travelling in some directions converge, while others diverge. The contours of a closed trapped surface, though, caused light rays to converge in any and every direction. The shape had powerful meaning, though Roger didn't fully understand its significance when he submitted the paper to the Royal Society.

Trapped surfaces were not front of mind as Roger hunted for singularities. But his greatest insights often came from the back burner.

His intuition told him singularities were real. If he could prove this before anyone else, it would change his career.

He had other reasons for hiding in the heart of a quasar. Joan was struggling, withdrawn, and tired. Roger didn't know what to do.

He knew how to perform intellectually. He excelled at finding brilliant solutions to difficult puzzles. But Joan wasn't a puzzle. He felt trapped, his life folding in on itself leaving no path for escape.

He spent long hours at Birkbeck College, working at his large chalkboard, covering it in curving, twisting diagrams. He had an unconscious habit of staring at the board, absently running the chalk over his lips, making them ghostly white. His notebooks from this time were filled with diagrams of converging light rays, twisted space-time, and speculation about the internal structure behind an event horizon.

Walking back from the Tube, Roger wandered in the woods, mulling the singularity problem and letting himself travel 2.5 billion light years away to C3 273. He lingered in intergalactic space as long as he could before entering his unhappy home.

He minimized his interactions with Joan, completing any necessary childcare before disappearing feet first through the living room floor. On weekends, family and guests would be startled by the trapdoor opening, Roger rising up out of the basement, immersed in a paper and oblivious to the conversations he interrupted.

One of those papers came from a team of Soviet theorists including Isaak Markovich Khalatnikov and Evgeny Lifshitz. It confirmed what most scientists believed: singularities were not a part of our physical universe. The circumstances that could create a singularity were indeed too improbable and specific ever to actually happen.

Kip Thorne, a Caltech physicist and future Nobelist, travelled regularly to Moscow in those years and remembers how important that paper was to the Russians. "They used perturbation theory techniques to tackle the problem. With a series of overlapping perturbations, they cobbled together what they regarded as a proof that generic initial data cannot produce a generic singularity," Thorne recalled. "They published this in a series of Russian papers a few years before Roger's theorem, and then they included it in the 1962 English edition of *The Classical Theory of Fields*. In some sense, this theorem was the crowning achievement of work on general relativity in their view."[2]

In the real universe, collapsing matter had enough chaotic motion to come swirling back out before reaching a point of infinite density. There had to be some other explanation for quasars.

Roger was sceptical. He intuitively sensed the Russians were making a false leap of logic, drawing broad conclusions from specific circumstances. Singularities might not form in the particular scenarios Khalatnikov and Lifshitz analysed, but he still wasn't convinced they were impossible.

The Russians were confident, and most of the scientific community agreed with them. But there was still the matter of a very small, very distant, very bright object out beyond Virgo, inexplicably dimming, brightening, and filling the universe with radiation.

In autumn 1964, Roger's old friend Ivor Robinson came to visit. Ivor still lived in Texas. The two hadn't seen each other in months, and Ivor had a trove of new stories and opinions to share. Ivor's loquaciousness was legendary among physicists.

"Never quite without irony, and a distinctly mocking wit, savouring his words like good wine, he loved to develop an intricate story or argument," Wolfgang Rindler said of him. "He was consummately verbal. No one who has heard Ivor talk will ever forget his inimitable style, his intensity and restlessness, his sonorous voice, his undiluted English accent, his brilliance, his natural dominance of any conversation."[3]

Roger and Ivor made an ideal pair: one always ready to talk, the other ready to listen. Their lopsided conversation was non-stop and wide-ranging, a steady stream of science, philosophy, politics, and culture. Though both were researching singularities, the subject did not come up during their morning walk or over lunch.

After sharing a meal, they chatted on the pavement near Birkbeck College, waiting for a break in traffic. Their conversation halted as the pair crossed.

In the brief silence of that street crossing, Roger's mind shot out again across billions of light years to the screaming hot mass of C3 273. He saw it as he had never seen it before. It might have begun as an amorphous dust cloud, a collapsing galaxy, or even a galaxy cluster. Its origins didn't matter to what happened next: gravitational collapse had taken over, pulling all the matter deeper and closer to the centre. Like a twirling figure skater pulling their arms in close to their body, the mass spun more and more quickly as it contracted.

As it shrank, it created a heat so intense that radiation blasted out on every wavelength in every direction. The smaller and faster it got, the brighter it glowed.

But there was more. Roger's imagination worked in four dimensions, not three. As the density grew, space and time warped more and more. He mentally mapped his chalkboard drawings and journal sketches onto the picture of C3 273, searching his mind for the point Khalatnikov claimed to have proven, where this cloud would explode back out again, slowing and cooling, continuing to obey the known laws of physics.

No such point existed. He at last saw how the collapse would continue unimpeded, how an event horizon would form, how gravity would become so concentrated that it would create a shape in space-time Roger knew well—a shape he had written about very recently: a closed trapped surface. At the very heart, space-time would wrench so dramatically, every direction would converge on every other. Just outside the event horizon, still-surviving matter would be churning furiously, emitting more light than all the stars in a galaxy—brightening and dimming with a power and speed beyond anything else in the known universe.

In that moment, he saw the thing almost nobody else believed possible: every light ray, every spatial vector, even the arrow of time itself converged on a single point. They all came to a halt. Density and space-time curvature went to infinity. Relativity broke down.

Not only was it possible, it was inevitable.

Back on the other side of the street, Roger picked up the conversation with Ivor and immediately forgot what he had been thinking about. They bid farewell, and Roger returned to the chalk clouds and stacks of paper in his office.

For the rest of the afternoon, Roger found himself in an unusually happy mood, which puzzled him. He had papers to mark, admin tasks to deal with, family worries waiting for him at home. But something felt different. Lighter. Exciting.

He reviewed his day, trying to pinpoint the source of his cheeriness. His mind returned to that silent moment crossing the street. It all came flooding back to him. He had solved the singularity problem. He had found a general solution to describe a gravitational collapse that had eluded him—and the world—for years.

He began writing down equations on his chalkboard, testing, editing, rearranging. The argument was still rough, but it all worked. With only some very general, easy-to-meet energy conditions, any sufficiently large mass could collapse to infinite density. Roger knew at that moment that singularities weren't some theoretical oddity—there had to be billions of them littering the cosmos.

At day's end, he exited the Tube, strode through Stanmore Park, passed Joan and the boys, and descended through the trap door to continue reading and thinking. He spent the following days and nights developing his singularity theorem, racing toward publication.

Within two months, he had begun giving talks on his singularity theorem. In mid-December, he submitted a paper to *Physical Review Letters*, which was published on January 18, 1965—just four months from the day he had crossed the street with Ivor Robinson and before the Royal Society published his paper on closed trapped surfaces.

"Gravitational Collapse and Space-Time Singularities" was only three pages long and dominated by geometry rather than algebra. "These were techniques that were not familiar to people. There was not much in the way of equations involved," Roger said.[4]

The Penrose singularity theorem was debated. Refuted. Contradicted. Misinterpreted.

American astronomer Bob Dicke slapped Roger on the back after one of his early talks. "You've done it! You've disproved general relativity," he said.[5] Roger didn't see it that way. He was still a relativist to his core. Einstein hadn't refuted Isaac Newton's laws, just refined them. Roger saw himself in the same mould—he'd merely identified the need for a new theory to supplement what already existed.

Roger's unusual geometric approach faced an uphill battle for acceptance. When Roger presented his theorem at an international congress on general relativity and gravity in London in 1965, most scientists still lined up to defend the Russians. "It was not very friendly. The Russians were pretty annoyed, and people were reluctant to admit they were mistaken," Roger recalled.[6] The conference ended with the debate unresolved.

The Russians did not welcome Roger's paper. "When Roger came out with his theorem, there was consternation in Moscow," Kip Thorne said. "They were very sceptical. On the other hand, in the West, there was huge scepticism of the Russian work because the level of rigour was nothing like the level of rigour of Roger's work."

Publicly, Khalatnikov's team cleaved to their publications. But privately, he and others began re-examining their calculations. Thorne happened to be there when they realized their mistake. "It was basically Belinski and Khalatnikov who found the error in their proof," he said. Their own calculations did, in fact, allow for a generic singularity to form.

"Lifshitz found it very embarrassing that they had made this error in what he had regarded as their most important work in the field. He wrote a paper confessing to the error. He was afraid that somebody in the West would find the error before they did and publish it."

He wanted the world to know they had found their own mistake. He knew Russian censors would not agree. Instead of submitting the paper through the usual channels, he gave it to Thorne to smuggle out of the country and submit to *Physical Review Letters* on their behalf. "I put the paper itself in with my own personal papers. I removed the title page, which had Lifshitz's name on it and put it elsewhere so the two were not tied together. I carried it like that back to the United States," Thorne said.[7]

After the Russians withdrew their argument, the Penrose singularity theorem began gaining traction. Roger's insight—years in the making yet created in the time it took to cross a London street—became a primary driving force in the near-miraculous advances in cosmology over the next six decades.

In late winter 1965, Dennis Sciama invited Roger to give a seminar on the singularity theorem at Cambridge University, where Sciama was still a professor. Roger had given versions of this talk many times at this point and was reasonably confident he would find a warm reception.

He spent hours creating overhead transparencies to accompany his lectures, drawing diagrams and sketches freehand with brightly coloured markers. His days creating visual games and impossible objects with his father served him well. His presentations had flare, with him flipping and overlaying transparencies, barely letting people absorb one image before turning it on its head to reveal something new. He understood the power of visual storytelling and the effectiveness of turning a talk into a performance.

In Cambridge he presented effortlessly and excitedly, laying one transparency down after another. There it all was—the lines bending in toward one another, the irrelevance of imperfections and variations in the surrounding material, the closed trapped surfaces, the irresistible gravitational collapse, and the inevitable, rumblingly beautiful creation of a singularity.

Most of the people in the room didn't share Roger's ability to imagine space and time in four dimensions. They tended to trust equations rather than pictures. But at this talk, everyone understood exactly what they were dealing with: the biggest advance in relativity since Einstein first published the theory. Roger had done more than explain quasars. He had revealed a major new truth underlying the reality of our universe. Whatever models people came up with from here on out, they were going to have to include singularities. And that meant they were going to have to include science that went beyond relativity.

A new chapter had begun. And Roger Penrose had taken his place as one of the greatest cosmologists of the twentieth century.

Roger packed up his transparencies as people filed out of the hall. Sciama stayed behind, accompanied by two men in their early twenties—a couple of his most promising graduate students whom he wanted to introduce to the man of the hour.

"Roger Penrose," Sciama said, indicating one of the two, a slight man with thick glasses, protruding ears, and a wide smile, "meet Stephen Hawking."

10

HAWKING AND PENROSE

Roger Penrose, Ted Newman, Wolfgang Rindler, Ivor Robinson, and Engelbert Schücking shared a view of the world few outside the physics community knew or understood.

They all had interests outside physics: Ted was an opera devotee. Ivor followed Israeli politics. Roger could get lost in the works of J. S. Bach and M. C. Escher—but physics bound them together and set them apart.

Within this elite, obsessive group, Roger had grown accustomed to having certain parts of the universe to himself. He could explore places none of his friends could reach.

Until he met Stephen Hawking.

Roger didn't have to translate his experiences or describe this world to Stephen. Stephen was already there. What's more, he didn't share Roger's aversion to equations; he could move effortlessly between formulas and geometry in ways Roger couldn't help but admire. "Stephen Hawking was different. He could make sense of the calculations all right. And he could certainly do geometrical visualization. He could do both," Roger said.[1]

Moments after being introduced, Roger and Stephen were deep in conversation. After another of Dennis Sciama's graduate students, George Ellis, joined them, the three men found a room with a blackboard and settled in for a long and pleasant night. Stephen and George had been collaborating on questions closely related to Roger's work, and the three easily riffed back and forth about the Penrose singularity theorem, taking turns sketching and diagramming at the board.

George and Roger both came from Quaker backgrounds. Nevertheless, it was Stephen who held Roger's attention and slightly unbalanced him with his overt talent and competitiveness.

Unbeknownst to Roger, Stephen had recently been diagnosed with amyotrophic lateral sclerosis (ALS), an incurable neurodegenerative disease. On average, people diagnosed with ALS died within fourteen months. Few survived more than five years. Stephen didn't have time for modesty or deference. Impending mortality sharpened his ambition.

Ego was not unusual among theoretical physicists. Roger, though, preferred Ted's sense of wonder and beauty, Ivor's joviality, and Wolfgang's gentle generosity to Stephen's urgency and arrogance.

Dennis asked Roger to serve as an external evaluator of Stephen's PhD thesis, which was partly based on the Penrose singularity theorem. Roger had shown where the forward-moving arrow of time reached its final destination in a singularity. Stephen flipped that idea, tracing time backwards through the cosmic past to its starting point: a *cosmological singularity* from which the universe emerged.

Roger had forced physicists to look beyond general relativity to understand black holes. Stephen required them to do the same for the creation of the universe. Stephen built on Roger's work so effectively, the Penrose singularity theorem swiftly morphed into the Penrose-Hawking singularity theorems.

It rankled Roger that they now shared credit. He felt his original theorem was a revolution and Stephen's work was a refinement.

"If it were stated correctly, I wouldn't mind. It's often stated incorrectly. Stated correctly, I had the theorem first. Stephen generalized it and applied it to different situations. He wrote three papers for the Royal Society. Only the last paper was jointly with me. I don't mind the two names coming together there," he said.

"I explained the general ideas of my techniques. Stephen picked up on them very quickly. He made a simple theorem that used my argument in the reverse direction of time. And then he developed techniques for general theorems about cosmological singularities. I was playing more with light rays, and he was looking more at timelines. The general techniques came from me. He developed them."[2]

Roger distinguished between scientists like Isaac Newton and Albert Einstein, who developed truly original ideas, and a larger pool of physicists who ran with them. He didn't think Stephen belonged in that first category but suspected he, himself, just might.

By then, Stephen's medical situation was public knowledge. Roger instinctively avoided public pettiness and strove not to begrudge Stephen his rapid ascent.

"He was certainly a young man in a hurry," he said.[3]

The Penrose-Hawking singularity theorems forced a reckoning with general relativity. They sowed new debates, theories, and confusion over why things are the way they are—and why there is anything at all. The sheer precariousness of physical reality was disconcerting. Very little would have to change about the laws of physics for atoms, stars, and human beings never to have existed.

The marvellous improbability of it all—a perfectly balanced universe in which atoms and particles coalesced into complex, self-aware creatures capable of understanding those very phenomena—suffused the work of scientists like Roger and Stephen with mystery and awe.

Hawking and Penrose drove a new wave of cosmological questions: Did our universe start out special, or did its extraordinariness emerge from some generic state? Does our physicist-producing species just happen to inhabit a region of the universe suited to organic life? Could constants and laws be different in other regions of the cosmos?

If the universe emerged from a singularity, quantum physics would have dominated its first moments. With its probabilistic, constantly fluctuating states, quantum mechanics might produce infinite cosmoses in a never-ending process of creation. Our reality might be one variant within an infinite multiverse.

Roger and Stephen continued collaborating, both for the satisfaction of trying to answer these huge questions and also, they hoped, for profit. They wrote a joint paper updating their singularity theorems to submit for an unusual and lucrative award, the Babson Gravity Prize.

American engineer and businessman Roger Babson created and funded this prize. In an essay titled "Gravity—Our Enemy No. 1," Babson wrote about the childhood trauma of his older sister drowning in a river near the Babson home in Gloucester, Massachusetts.

> Yes, they say she was "drowned," but the fact is that…she was unable to fight Gravity which came up and seized her like a dragon and brought her to the bottom. There she smothered and died from lack of oxygen.
>
> Gradually I found that "old man Gravity" is not only directly responsible for millions of deaths each year, but also for millions of accidents….Broken hips and other broken bones as well as numerous circulatory, intestinal, and other internal troubles are directly due to the people's inability to counteract Gravity at a critical moment.[4]

Babson received an engineering degree from the Massachusetts Institute of Technology in 1898 and earned a fortune by patenting the first coin-operated parking meter. He shifted to investment banking, where, despite delusional ideas about gravity governing the rise and fall of the stock market, he became a multimillionaire.

He channelled his wealth into funding scientists who he hoped would create "anti-gravity devices, partial insulators, reflectors, or absorbers of gravity"—weapons and armour to tame this deadly monster.

"He was very rich and had bad feet," Roger recalled. "He thought, if somebody could invent some kind of cell to lift you off your feet and shield you from gravity, then he would benefit from that invention. We all thought it was pretty crazy. Nevertheless, Stephen and I decided we would put in for this prize."

They thought they had a good shot of winning the $4,000 purse: Babson's foundation might look kindly on a paper focused on situations where gravity goes haywire. In their paper "On Gravitational Collapse and Cosmology," they identified three such scenarios.

We present a new theorem on space-time singularities. On the basis of the Einstein (or Brans-Dicke) theory,...we show (essentially from the property that gravitation is always attractive) that singularities will occur if there exists either a compact space like a hyper surface or a closed trapped surface or a point whose past light-cone starts converging again. The first condition would be satisfied by any spatially closed universe, the second by a collapsing star, and the third by the observable portion of our actual universe.[5]

Unlike other forces, gravity always attracts, never repels. This gives it a unique inevitability. When every other force burns out or self-cancels, gravity will be waiting. It's weaker than magnetism and the nuclear forces but has an unmatched inexorability.

When a star of greater than 1.5 times the solar mass exhausts its nuclear fuel and cools, the pressure forces are insufficient to reset the gravitational attraction. What then would happen to such a star? Would it collapse to some sort of singularity or would it happen that the smallest departure from spherical symmetry would cause different parts not to fall exactly towards the centre and so miss each other.

In the reverse direction in time a similar question arises in respect of the whole Universe: Was there a singularity in the past or did the Universe manage somehow to pass smoothly from a contracting phase to the present expansion?...We shall present a new and very general theorem which shows that singularities would be expected both in the collapse of a star and at the beginning of the expansion of the Universe....In such a region, the presence of any matter would cause infinite curvature.[6]

Stephen and Roger, two powerhouse collaborators, rarely met in person. They spoke by phone and wrote independently. "He wrote a version and then I rewrote it and put in a lot more stuff," Roger recalled. "Each one would send it to the other, and say, 'Any suggestions?' Stephen put in an appendix which was more or less his about whether the conditions for this theorem were established by observations of the Cosmic Microwave Background. I took his word for it basically."

Their paper took second place, behind a paper by Harvard University astrophysicist David Layzer, who argued the universe began not with a singularity but with a sphere of nuclear liquid with a temperature close to absolute zero. Shortly after his win, Layzer's cold Big Bang theory was soundly disproved.

In April 1969, Stephen and Roger submitted a second coauthored paper with similar content to the *Proceedings of the Royal Society of London*. When this paper, "The Singularities of Gravitational Collapse and Cosmology," was reproduced in Roger's collective works, he subtly tried to reassert sole ownership of the original idea.

"Following my demonstration of the singularity theorem... Stephen Hawking produced a series of papers containing technical mathematical advances which effectively generalized this result so that the implication of the presence of space-time singularities in generic situations would apply also to the cosmological context."[7]

He created. Stephen elaborated.

Roger's reputation was well established before he met Stephen. Yet somehow Stephen rocketed past him toward mainstream celebrity. ALS was taking visible toll on his body, but he had already outlived his original prognosis. His continued productivity and brilliance made him an icon of the triumph of the mind over the body. People who didn't understand the nuances of the Big Bang easily absorbed the power of Stephen Hawking's personal story.

Illness also gave Stephen powerful motivation to pursue mainstream success: he needed money to cover medical and physical care.

The two men remained collegial competitors.

Many scientists shared Roger's mixed feelings about Stephen Hawking. "Stephen was a very difficult person," said Paul Tod, an Oxford mathematician who knew Stephen and Roger well. "There were all these stories that he would run your foot over with his wheelchair if you stepped out line. But Roger always adored him. He wouldn't hear a word against him in the news."[8]

Privately, though, he was often frustrated with Hawking.

In 1971, Roger travelled to Cambridge to give a talk on singularities. He met Stephen afterward to run a new idea by him. "After I'd given my talk, I went to an office, a room where you could meet with people with a blackboard. I talked to Stephen about a result which had to do with event horizons. I described to him how the surface area of the event horizon of a black hole has to increase over time. As you go into the future, it can never get smaller," he said. "He listened to me and didn't say anything."

The idea related both to his developing models of black hole behaviour and also to bigger questions of why time appears to move in a particular direction and why the past and the future seem so different.

Time connects to the second law of thermodynamics, which stipulates that disorder and randomness—*entropy*—always increase. This unidirectional evolution creates the perception of an arrow of time. The future is distinguishable from the past by its increased entropy. Roger realized the future also had larger black hole event horizons than the past—another temporal asymmetry.

After more than an hour, Roger still wasn't sure he had explained the idea well enough to Stephen, but it was late, and he returned to his hotel.

Stephen phoned me the next morning. He said he had a new result, about black holes colliding, and finding limitations on the amount of radiation they could release using the surface area. It was basically the idea that I'd been talking to him the night before.

He said, "Your idea you told me about yesterday applies to the situation where two black holes collide and make a bigger black hole. The future black hole must be equal to or greater than the sum of the areas of the previous black holes."

Stephen had done it again—absorbed one of Roger's ideas and taken it further than Roger himself had imagined. Stephen went to work on a new paper. Roger remembered,

I don't think he mentioned to me in the paper. He talked about the surface area theorem being a most wonderful result, which he thought of getting out of bed in the morning. I think the situation was more or less as follows: I talked to him about this the night before. He didn't fully understand what I said. The next morning, he was getting out of bed, and he thought it through again without realising quite what I had said to him. Rather than going back and saying, "Is this what you said? Oh yes? I'm using that result," he used it in this other context, which is not a context I had thought of. Certainly, I give him credit for that because it was a new result. I couldn't quite make out, Was he being dishonest? Would he really not remember what I had said? He understood enough of what I said that if he wasn't putting himself forward, he would have said, "Oh gosh, this is what Penrose was saying." I just felt that there should have been a bit of recognition about that later.[9]

Though Roger insisted "it wasn't hugely important," the incident stayed with him for another fifty years. "It's just that somehow this particular thing was something he considered to be such a big step in what he was doing—without mentioning that it was based on something that I told him."[10]

Roger didn't like having to be wary around colleagues. He resolved to stick with collaborators like Ted and Wolfgang who appreciated physics for its own sake rather than for how it could build one's reputation.

Roger's and Stephen's public reputations remained enmeshed, even when each subsequently rejected major aspects of their original work. Stephen discovered a process by which black holes could radiate energy, causing their surface areas to shrink. *Hawking radiation* was such a powerful contribution to the field, even Roger praised it without caveats.

Meanwhile, Roger came to doubt Stephen's story of the universe's origin. His dissatisfaction with the cosmological singularity would fester for years, widening their divide.

In the late 1960s, a school of thought known as *string theory* was gaining traction with particle physicists working to reconcile general relativity with quantum mechanics. String theory treated electrons, photons, and other fundamental particles not as point-like objects but as "one-dimensional strings." These strings' vibrational patterns manifested as properties like mass, charge, and spin—the defining properties of subatomic particles. String theory predicted the existence of a *graviton*, a quantum particle carrying gravitational force. It seemed at last to unify physics by creating a theory of quantum gravity.

Roger found string theory aesthetically beautiful and conceptually absurd. In part, he objected to some of its more outlandish predictions.

String theory, for instance, required more than the usual three spatial and one temporal dimensions. Different versions used ten, eleven, or twenty-six dimensions, most of which were *compactified*, or curled in on themselves so tightly as to be imperceptible by human senses and scientific instruments. He used compactification in his own work but saw no justification to apply it to extra dimensions.

Roger hated predictions like these—untestable, unfalsifiable assertions with no particular reason to be true other than they satisfied certain mathematical equations.

On one of his trips to the United States, he attended a meeting of physicists at the Stevens Institute of Technology in Hoboken, New Jersey. In the parking lot outside the conference building, he sat in the passenger seat of a car belonging to his friend and colleague Jim Peebles (who later won the 2019 Nobel Prize in Physics—just a year before Roger).

Roger wondered out loud why he resisted ideas like string theory so strongly. Was string theory really that much stranger or more counterintuitive than relativity or quantum theory (or twistor theory for that matter)? Why was he so unwilling to even entertain ideas like extra dimensions if they allowed for a working theory of everything from the smallest particle to the largest galaxy?

"You resist because the universe isn't like that," Peebles told him.

These words struck something deep in Roger. They resolved an inner debate. Despite its chaos and hidden secrets, the universe was at heart just what it appeared to be. It wasn't necessary to add extra dimensions or particles or forces to make sense of it. Roger left the Hoboken parking lot with new clarity. He would try to understand the universe as it presented itself—as it was. He would reject theories that required mathematically conjured add-ons to make sense. If a theory described a universe other than the one he lived in, he had no patience for it.

Stephen Hawking reacted differently. Roger speculated that Stephen's openness to string theory might have something to do with his health. "I think that he believed his time was limited, and if he was going to find a route into the theory of everything or whatever they used to call it in those days, he would have to follow the string theorists and do what they were doing," he said. "I mean, to me, that's not a good enough argument. But I can see in a sense why he might put his money on string theory if he felt his time was limited."[11]

Stephen was no Roger Babson—so consumed with mortality and the frailty of human life as to devote his efforts to crackpot theories and doomed ideas. His work fell well within mainstream theoretical physics. But Roger had as little time for those ideas as he did for Babson's dreams of a gravity insulator.

Jim Peebles helped him quell his self-doubt, reassuring him he wasn't going off the rails like Babson. Roger was rejecting the mainstream, not the other way around. He was speaking truth to power, not shouting at the clouds. He had been right about singularities, even when world said he was wrong. And if most of the physics world was off chasing strings, that only left him more room to pursue his own theories.

11

JUDITH DANIELS

In 1970, Roger reluctantly wound down a series of visiting professorships and informal stops at Yeshiva, Syracuse, Cornell, and Princeton universities. This excursion nurtured him emotionally and intellectually. In particular, he had extended time with Ted Newman, exploring spinors and relativity, trying to decipher Newman-Penrose constants, and going on hikes and camping trips, immersed in the natural world they were trying to understand.

The success of his singularity theorem and the potential for twistor theory assured him, at age forty, that his most important work still lay ahead. It was an exciting time to be a physicist, with new discoveries happening everywhere at once: NASA astronauts had landed on the moon and were planning to go back. Experimental physicists using electron scattering guns had revealed the shape of atomic nuclei.[1] In a mineshaft 1.4 kilometres below Lead, South Dakota, scientists had built a neutrino trap to capture and study these elusive fundamental particles.[2] Orbiting telescopes were making the first ultraviolet scans, promising new information about how stars, galaxies, and the universe itself formed.[3]

Material scientists had found metallic alloys that behaved like malleable plastics.[4] Space probes were analysing the atmospheres of Mars and Venus.[5] Atomic physicists were exploring the idea that protons weren't actually the indivisible, fundamental particles they had once thought but instead were made up of even smaller components. Everywhere, people were looking farther and deeper, out to the most distant galaxies and into the tiniest particles, refining and redefining basic knowledge of the physical world.

As a theorist, Roger craved getting underneath each new observation to find generalized rules that explained them. Even taken together, general relativity and quantum mechanics couldn't answer the questions that nagged him in middle of the night: Why were there three spatial dimensions? Why not two or eleven? Why can we remember the past and not the future? Why did the universe have an unbreakable speed limit? Why does gravity's pull decrease proportionately to the square of the distance between two objects? Why were there four forces, seventeen fundamental particles, and ninety-two naturally occurring elements? Every constant, law, and property had to come from somewhere. There had to be an underlying structure that could explain everything.

He was convinced the problem was solvable. He increasingly believed he might be the one to solve it.

He had taken Jim Peebles's advice to heart: focus on explaining the universe as it *actually* is—simple, consistent, and knowable.

Since his 1963 epiphany on the road to Dallas, Roger remained convinced that twistor theory could reconcile relativity and quantum mechanics. He was disappointed so few colleagues shared his passion for twistors but also liked having them to himself. If his ideas panned out, the results would be his alone.

He was also excited that his sons had started asking him questions about mathematics. Toby and Christopher were about the same age Roger had been when he discovered the sundial in the Jungle. Roger and Joan's third son, Eric, was just four years old—still too young to really join the conversation.

"What things have more than three dimensions?" Toby asked one morning. Roger felt the significance of the moment—he could lift his sons up the way his father had done for him and introduce them to the vast world awaiting them beyond everyday experience. He would guide them into worlds of one, two, three, and four dimensions, giving them their first glimpses of the true nature of space-time.

"I feel it might really be possible to get over some of the basic ideas of relativity to quite young children," he wrote.[6] He had more capacity than Lionel to chase his children around and make them laugh, but he still gravitated toward science as his primary form of bonding.

Back in England, his optimism quickly dissipated. At Stanmore, peace and quiet were less achievable than ever, even through the living room escape hatch. At Birkbeck College, he felt restless and underworked.[7]

Roger's sister Shirley also lived in London. Joan admired Shirley so much she felt inadequate in comparison. "Shirley went to several horrible private schools. She stuck to her guns, and she got a medical degree, which was not easy for a woman," Joan said. "She was a very determined girl. She worked hard and showed a lot of self-belief—more so than I have done. Shirley put me to shame really. I was majoring in men, and she was majoring in medicine."

Roger and Joan periodically dined with Shirley and her fiancé, a University College London professor of medicine named Humphrey Hodgson. At one dinner, Shirley mentioned to Roger that her childhood friend Judith Daniels now studied mathematics at John Cass College. Roger had last seen Judith years earlier when she was still a teenager. He recalled how simple and joyous that friendship had been when so much else felt so awkward. He could use some of that feeling in his life again.

He rang Judith and invited her to dinner. They met at an Italian restaurant near his office. Not knowing what to expect, he was surprised by everything.[8]

She was no longer a skinny, rambunctious teenager. Her hair had grown longer, and she no longer tied it back. It framed her face, in which Roger saw the same inquisitive eyes, the same warm smile, the same joy he had appreciated when they were younger.

She laughed when she greeted him with the same delighted snort that used to uplift the entire Penrose family. She still exhibited all the vivaciousness he remembered. And now she was an adult human being who spoke the language of mathematics—his language. He was stunned to discover she not only understood his accomplishments but was impressed by them.

His spirits lifted when he talked to her. He felt an optimism he hadn't realized he'd been craving. He unburdened about spinors and twistors, his hunch that he was on the trail of the complex geometry that could lead him straight to the fundamental structure of the universe.

She asked perceptive questions and devoured the answers.

He left the restaurant elated. He could barely understand what had just happened. He became sceptical of his own reactions, fearing misery had put him in a "susceptible frame of mind." Was this actual happiness or "just some mental construction?"[9]

He told Joan he'd had dinner with Judith.

"But suppose you fall in love with her?" Joan said.

Roger didn't answer. But he thought to himself, "Yes, that would be rather nice."[10]

They met regularly for lunch or dinner and talked about spinors, twistors, and space-time.

Judith confessed that she had had a crush on Roger when she was a teenager and "worshipped him from afar." She said she'd felt heartbroken when she heard he was getting married.[11]

Adulthood changed her perspective. When she was nineteen, a few years after she had lost contact with Roger, she married a man named David Tennant. Their marriage was devastating and brief.

"She longed to have children. When she discovered during her first marriage that she was not able to, it was a great sadness for her," her sister Helen recalled. "She was always very loving. I think she saw herself in this loving role, this maternal role. After that marriage ended, she didn't want to commit herself to anybody because she felt she could not fulfil that role."[12]

Roger didn't fully appreciate Judith's transformation. She still laughed and sparkled like she had as a child. She seemed an adult version of the girl he once knew.

It was easier for Judith to see how Roger had changed. He was no longer her school friend's older brother but a respected, somewhat famous mathematical physicist. He spoke all over the world. Iconic optical illusions and influential physics theories bore his name.

Yet he solicited her feedback on his ideas and asked her, an undergraduate, to read drafts of his papers and talks. He talked her through his scribbled equations and diagrams and was overjoyed at her responses. He introduced her to a hidden world where the shape of things revealed the underlying workings of the universe.

She was fascinated and flattered.

He found himself relying on her more and more for validation and motivation. Merely being around her changed who he was, made him feel more creative and driven. New ideas came more often and more easily. He didn't see her as a collaborator but felt she had a mysterious power to draw ideas out of him.

In July 1971, Roger and Joan planned an extended trip to North America, flying first to Detroit to visit Joan's family, then to Ontario for a few weeks at Blink Bonnie, and finally to Dallas, where Roger had planned a writing retreat with Wolfgang Rindler at the University of Texas.

Spinors and Space-Time still wasn't finished and had been proceeding painfully slowly. Roger's theories constantly evolved. Both men got caught up in other work for months at a time. When they returned to the manuscript, they often felt like throwing everything out and starting again. They were hoping an extended Rindlerizing session would get them back on track.

Dark though his memories of Texas were, Roger was glad for an excuse to take a break from England and his family. He told Judith he'd send her chapter drafts for feedback while he was away. He sent his first letter from Joan's parents' house in Detroit.

"We're staying with my in-laws for a few days, but I have even managed to get a little work done here, which is unusual," he wrote. "All in all and with a newly found feeling of optimism (which I owe to you, of course) things do not seem to be at all bad at the moment."[13]

Roger carried on with his family to Bayfield and Blink Bonnie. The cottage, which they now owned, was still rustic but vastly more comfortable than when they first visited a decade earlier. Christopher, Toby, and Eric explored the quiet streets, rugged cliffs, and stony beaches along Lake Huron. In theory, this leg of the journey was a family vacation, but Roger wrote through the night when Joan and the boys were asleep, retiring at 6:00 a.m. and rising around noon.[14]

Roger raced the boys around the cottage and wrestled with them on the front lawn. The whole family loved it there, and the change of scenery helped lift the mood.

From Bayfield Roger mailed carbon copies of a handwritten draft of the second chapter of *Spinors and Space-Time* to Judith with a note hoping they would afford her "some light entertainment."

"I'm thinking of this book writing as really just an enormously long (and moderately boring) letter to you. I hope you don't mind," he told her. "At least it makes the writing more palatable (for me that is)."[15]

He told her not to worry that the subject matter was not her area of expertise.

"The question is, is it tolerably readable? Is it tedious? Is it repetitive? Is it at all interesting? Is it moderately well motivated? Does it make any sense? Is it on a level comparable with Chapter 1? Or is it really a different book? I really have no idea, since I'm so close to it, so an outside view would greatly help, especially one with not too many preconceived ideas about it all."[16]

He rubbed his chin, which he hadn't shaved since landing in North America.

"I now have over 3 weeks growth of beard. I'm just toying with the idea of keeping it. But probably it's not worth the extra 3 hours of searching for drugs at customs when we re-enter the U.S. Do you have views on beards?"

Considering his own appearance led him to think about hers.

"Just one final thing. Do you have a moderately recent photograph of yourself that you could spare me? (Also an ancient one if you have one. That would be nice too. Better still, put yourself in the envelope—not C.O.D. though!) Things have been OK so far, but it would be nice to have something extra to sustain me though the book writing in Dallas."[17]

Judith wrote back from London a week after receiving the draft, guilty that she hadn't yet read it.

"I am at a loss for words," she wrote. "It sounds silly to say I feel 'honoured,' but that's the way of describing it. I don't know—I'm glad, flattered, everything, that you should write to me, and for me. And right now I feel unworthy too."

She was working at an administrative job at London's Medical Research Council. She was also cooking and caring for an elderly couple living in the flat below hers. "They're 96 and 89 and, though both mentally lively, of course, it's terribly difficult and slow for either of them to do even the simplest food at home. They have meals-on-wheels at lunchtime, but they don't operate at weekends."

She promised to find time to read the draft soon, but reminded Roger, "I'm not a typical reader in that my background knowledge, if any, comes mainly from you and...I'm terribly biased—because I know you."

Their connection electrified them both. Roger was euphoric. Judith was more careful about separating fantasy from reality.

"My mind strays into tabu territories and daydreams. Do you realize how many hours I spent between the ages of 10 and 15 just imagining that I was grown up and that you would feel about me as you seem, from your letters, to feel now?" she wrote. "But it's not a bit like the dream, is it? Try and understand. Because I worshipped you, because I still believe you are my 'ideal,' you are the last person in the world with whom I could have an 'affair.' Please be careful that it doesn't develop into that."[18]

She said she liked beards and didn't have a recent photo but would try to find an old one.

Roger paid no heed to Judith's worries about feedback and affairs.

"I am most grateful for you to be so explicit about the difficulties you have with objectivity, etc.," he wrote. "Please don't feel under pressure to go through all that MS if you don't feel like doing it. As you no doubt realize, sending it to you was to some extent a psychological device (for me)."[19]

He didn't need her mathematical expertise or her editorial feedback. He had Wolfgang and other colleagues for that. He sought something else from her. Her words, her laugh, her smile—all became essential tools for his scientific endeavours.

"The psychological need is less now that Wolfgang is going over the MS and it's becoming a joint project again," Roger wrote to Judith. "But I still find writing this stuff a terrible bind. (It was 9 years ago when we embarked on this project—before you were married!) So please try to dig up some kind of photo. It doesn't have to be very recent. Anything up to 20 years old would do. (Not too much older than that, though please!)"[20]

He acknowledged her attempt to set boundaries but did not accept it.

"I should just say that my feelings towards you are very far from superficial—I think 'deep' is the word rather than 'intense,' having their roots very much in the past. . . . I don't think I realized the extent of it before recently," he wrote. "I see many problems and hardly any solutions. But I think that optimism (even fantasy) is an essential part of progress.... I think one has to keep an open mind about the nature of a solution to a problem. At least that's true in mathematics. Maybe it's true in real life too."[21]

Judith wrote again before Roger's latest letter had reached her. Guilt had gotten the better of her, and she'd devoted five hours to the manuscript. She tentatively noted down her thoughts and comments. ("I'm sure you'll find them disappointing after all this suspense and delay!")[22]

She felt badly that she couldn't check his proofs more carefully and had to "look things up and so on."[23] She felt it really did read like a letter written directly to her.

"I made notes as I went along and five times I wrote down a question or suggestion which I had to cross out because it was answered or complied with in the following paragraph or page!" she wrote. "Although I'd be more than flattered, I don't *want* to believe you're writing it for me. That's difficult to explain but it's to do with how I feel about you and needing to believe in the much more universal importance of what you're doing."[24]

She worried Roger invested too much in the idea of writing the book for her. He was enthralled with the possibility she held the key to unlock his ideas. She felt the pressure and also worried the book would suffer. She knew, even if he didn't, that the actual audience for *Spinors and Space-Time* was an international community of mathematical physicists and relativists, not an undergraduate in another field. Nevertheless, she tried to help, providing detailed feedback on readability and gamely questioning Roger's reasoning.

Roger was particularly concerned about a new type of notation he had invented that made it easier for him to reason and calculate with spinors. Spinor notation was typically written algebraically—using formulas "encumbered with numerous small indices" and letters and symbols from multiple alphabets to signify various concepts.[25] To avoid the ordeal of paragraph-length equations, Roger invented a visual notation system that distilled algebraic equations into simple line drawings. It made his life easier, but he was uncertain other researchers would embrace it. Most of his colleagues were fine with equations. And he knew publishers would hate his notation because it couldn't be printed in any standardized way.[26]

Judith encouraged him. "I'm biased in favour of your notation....I didn't find it that difficult to absorb. And surely most of your readers will see the advantages (very well laid out to them) and be willing to change their methods. After all, they will, if they've worked so much with the old system, know all its disadvantages. Quite honestly, apart from the fact that its more useful, I find your notation mathematically more pleasing because it is more general."

With editorial feedback out of the way, she took pen, paper, and her cats to bed to continue writing to him about more personal news—her unhappiness with her job, a row she had with a male friend who accused her of being "incapable of love," the question of whether a therapist could help her deal with her persistent sense of guilt.[27]

"I think I can sort it out myself. I'm very wary of mucking about with 'natural' mental processes. I'll admit it's mucked up my 'sex life' but, even if analysis could 'cure' that, would it be a good thing? It's an easy and commonly accepted assumption that one is emotionally healthy when happily paired off but...I think I need a period on my own," she wrote. "I've hurt and/or disappointed several people and it's brought neither myself nor them any joy."[28]

When they left Bayfield, Joan, Christopher, Toby, and Eric flew back to London. Roger carried on to Dallas, where Wolfgang had secured him a small apartment near the airport. The flat was close and dark, the heat oppressive. Departing planes buzzed low overhead, creating a deafening roar whether the windows were open or shut.

Within a few days, Roger had covered the tables, chairs, and floor with sheafs of handwritten drafts. Surveying the room, he came close to despair thinking about the volume of work ahead. He might never have enough space or time.

He pushed work aside and reread Judith's latest letter. He discovered a vent on the small air conditioner that let in a bit more air. On a blank piece of paper, he wrote Judith's name over and over to cheer himself up.[29]

His spirits lifted a little. He sat down to work but discovered he was writing Judith instead. "I'm not sure that I ought to be writing this letter," he began, "but I do feel the need to set down my thoughts and emotions rather more fully than I have previously."

He dropped all pretense about wanting her feedback on his work. He told her how lonely he had been when they first reconnected, how Joan wanted to be rid of him. Soon he'd be separated from her and "even from my three children whom I deeply love and who, I am sure, get something irreplaceable from me in return."

In Judith, he at last saw a ray illuminating a path of escape from his marriage. "When I think back on those seemingly all too few occasions on which I saw you, you appear like a beacon of light to me. You were young, yes, but you had a quality that I have never found in anyone else before or since—a quality that was somehow able to brighten my life."

He was consumed with her ability to brighten a room, enliven a conversation, and lift his spirits.

I don't really know why I'm going into all this, but I just feel the need to say it. Perhaps it puts my present feelings towards you a little more in perspective. But does it mean that my present picture of you is a reconstruction out of a number of old dusty images? Perhaps to some extent it is. I don't really have too complete a picture of you at the present time. But no matter how I try to fill in the unknown details about you, it seems to make no important difference to the way that I feel about you.[30]

He was just as thrilled about the way he thought she felt about him and especially about his work.

"Having you read and actually understand some stuff I have written really does something for me....It is such a contrast with what I have been used to—the constant feeling of guilt, that what I am doing is not appropriately directed towards immediate gain, etc. It is so refreshing that you should appreciate that these things can be done for their own sake, that there can be real art and beauty in it."

When he saw himself through Judith's eyes, rather than through Joan's or Lionel's, he could finally believe he was doing something important. She could step into his world, move among the complex light rays, and walk with him through a universe of elegant, beautiful mathematics.

He worked through the autumn on *Spinors and Space-Time*, breaking up the work with trips to Austin, Boston, Pittsburgh, and elsewhere to give talks on black holes, the philosophy of science, and twistor theory.

Writing to Judith made him happier. His life was simply better for having her in it.

Judith found their correspondence more stressful. His drafts were increasingly challenging, and she assumed this reflected her own shortcomings rather than those of the text. "I haven't spent long on it, but I found I had to keep referring back. And I'm afraid it didn't seem to read as well—hadn't got the same flow. If it is a letter to me, you greatly over-estimate my intelligence and retention!" she wrote.[31]

She promised to spend more time with the book but struggled with other demands, including worrying about the elderly couple she helped care for.

Mr. Warrin's been rather ill. He gets terribly depressed sometimes and just wants to die. I seem to be the only person who can get through to him when he's like that. The trouble is that so many people still can't see depression as an illness. They say, "Well I feel depressed sometimes, but I pull myself out of it." They don't understand that he can no more pull himself out of it than he could cure himself of cancer. He doesn't eat because he really believes he's just a nuisance to anybody and he genuinely wants to die.

• • •

Judith may not have known about Joan's depression, and Roger didn't connect her commentary back to his wife. He was too focused on Judith. He dispensed with the charade of asking for editorial feedback. With each letter he wrote, he grew less apprehensive and more direct. With each letter he received, he felt more encouraged. He worried she might only interact with him out of pity—that she was taking care of him the way she did the Warrins. He worried he was being selfish and "not sufficiently taking into account any effect it might have on you."

He had no specific plan for leaving Joan. Furthermore, Judith was clear she didn't want a romantic relationship with him. Still, he interpreted her willingness to keep writing him as a sign of hope. He set about dismantling her objections, starting with the relationship they'd had when she was a child.

> The father-daughter aspect of our relationship undoubtedly has some importance in my feelings towards you. Perhaps it's responsible, in part, for the great depth and permanence of these feelings. But by the very nature of our situation, I would have expected this (F-D) aspect to have assumed rather greater proportions with you than with me. You were, after all, only 6 when I first knew you. I'm only mentioning this because I think our situation is complicated enough anyway that all factors must be taken into account and sorted out as best as possible if we are to be able [to] find something workable.[32]

He assumed they were both trying to find "something workable."

In early October, Roger wrapped up his stay in Dallas, the book far from complete. He travelled to Pittsburgh to visit Ted and give a talk, then returned to England, his family, and teaching undergraduate classes on relativity and on tensors and differential forms at Birkbeck. He kept searching for a solution to the F-D aspect of his relationship problems with Judith.

"I don't want to suggest that I don't agree with your very reasonable fears...but it really does seem to me that, at least from my own point of view, the only solution (at least for the time being) to the very real problem that I have in regard of my feelings towards you must be to channel in a constructive direction," he wrote. "At least we can try. At least we *must* try."

12

TRIESTE

In 1972, the international physics community had sufficient talent—and ego—to ensure no consensus on who was the smartest person in the room. But within that elite group, many people recognized Roger Penrose's highly unusual abilities.

"Roger's math had something magic about it," recalled Rainer Sachs. Sachs, the German-born cosmologist who worked down the hall from Roger in Austin, Texas, at the time of Kennedy's assassination, had watched Roger's reputation grow through his collaborations with Stephen Hawking. "Comparing the two, I always felt Hawking's calculations were comparatively straightforward—had I been smarter, better educated, more focused, and harder working I might have come up with some of them. Roger's were completely alien: Never could I have started one of his calculations, which always seemed to begin with an epiphany. Hawking was a human genius. Penrose's insights seem to stem from some superhuman life form."[1]

Many relativists had a powerful feel for formal math: errors in calculations leapt out at them the way off-key notes rankle a musician's ear. Not Roger. Equations required too much mental labour and restricted his creativity. His "magic" came from the shape of things. He preferred to run his fingers along the curves and twists of space and time and find in those graceful lines the story of how every particle, force, and phenomenon acquired its properties.

Complex geometry, with its poetic mix of real and imaginary numbers, held one key to unlock the relationship between general relativity and quantum mechanics. Conformal geometry, with its scale invariance and preserved angles, contained another. Together, they offered passage into the mathematical world "hiding behind the real world."[2]

Geometry had as much—or more—material reality as a piece of chalk or a wine bottle. Roger could manipulate and shift his mathematical constructions with an ease that seemed otherworldly to those around him.

"I know no one who even approaches him in intuitive feel for the content of alternative formulations of geometry, and no one who comes close in his lectures and writings [to] bringing such a wealth of insights to the service of his colleagues. I count him a unique asset to the world of learning," John Wheeler wrote of him in 1972.[3] Wheeler, a titan of theoretical physics, was influential not only for his own science but also for his ability to identify and encourage talent in those around him. He could name fewer than five scientists anywhere in the world who could come close to equalling Roger's accomplishments and potential.[4]

That year, Roger was under consideration for election to the Royal Society, the United Kingdom's prestigious national academy of science. "If the Royal Society wants the most distinguished man easily named in the field of mathematical physics it should lose no time in electing Roger Penrose," Wheeler told the selection committee.[5]

Wheeler was trying to lure Roger from London to Princeton, but Roger's memories of Texas put him off moving to America permanently, much to Wheeler's disappointment. "I must say it has taken me some time to become adjusted to the sad thought that there are factors about this country which rule it out for you in the years immediately ahead," he wrote to Roger. "If outlooks were to change...I personally would hope that somehow or other it might be possible to renew the proposal which we have just been discussing back and forth across the Atlantic so seriously."[6]

Roger's public reputation was growing alongside his academic status. His singularity theorem seeped into the public imagination, though under another name: a science journalist named Anne Ewing used the term *black hole* in a 1964 feature for *Science News*.[7] Wheeler embraced the term a few years later, making it famous. It captured people's imagination in a way *singularity* and *event horizon* never had.

Roger found himself increasingly in demand as a public expert on one of the hottest pop science subjects of the decade. With thoughts of Albert Einstein, Robert Oppenheimer, Fred Hoyle, and Stephen Hawking, Roger believed he might have a future as a celebrity scientist.

Scientific American invited him to write a feature outlining the latest research in black holes. He embraced the chance to introduce readers to an unseen world, strange and spectacular beyond imagining.

In what was becoming his signature style, he edged into his subject gradually, starting with something familiar to readers: our own sun. In about five billion years, he explained, the sun will run out of fuel, at which point it will explode, becoming a *red giant*, a star so large it will swamp the orbits of Mercury, Venus, and likely Earth. The sun will then collapse into a *white dwarf*, a small, dim, very dense, and nearly dead star, about the size of Earth.

Stars larger than our sun die a different death: a bigger explosion, followed by a more complete collapse, he wrote. A star twice the mass of our sun wouldn't stop collapsing when it hit white dwarf density. Gravity would pull it further inward, melding individual atomic nuclei together and compacting the remains, until it finally came to rest as a *neutron star*, a dark, dense ball less than 1/700th the radius of a white dwarf, with a density 100 million times as great. "A ping-pong ball filled with material from a neutron star would have the mass of...a minor planet 118 miles across," Roger wrote.[8]

At last, he led readers to the actual subject of the piece, the rips in the fabric of space-time caused by the death throes of even larger stars. "It is not possible to obtain any further stable equilibrium state.

The gravitational effects become so overwhelming that they dominate everything else.... We are led to a picture so strange that even a neutron star seems commonplace by comparison." All the matter of the star would disappear beyond an event horizon, a spherical one-way exit out of the observable universe. Any light or matter that crossed that boundary could never return to normal space-time.

The physics of black holes could not have been more perfectly suited to Roger's sensibilities and talents. As a large star collapses, its physical properties are stripped away. It no longer has colour or brightness. It isn't made of any particular elements from the periodic table. Its internal structure can't be observed. It has mass, spin, and an electrical charge, and no other properties. They aren't so much made out of "stuff" as they are a region of space-time that extreme gravity has twisted beyond comprehension. A black hole is, in essence, a shape. Like the wedge of a sundial. Like a polyhedron. Like an impossible triangle.

In *Scientific American*, he went on to describe even larger black holes with millions or billions of times the mass of our sun—the kind of supermassive object physicists theorized might inhabit the centre of our own Milky Way galaxy. At such sizes, the boundary between the familiar universe and the interior of a black hole occupies a large, gradual transitional region.

Roger described an astronaut tumbling into a black hole 100 million times the mass of our sun. The event horizon would form a sphere millions of kilometres in diameter. Nearer the central singularity, density and gravity would approach infinity, creating tidal forces strong enough to rip a person apart. But the outer regions, just within the event horizon, would be surprisingly habitable.

"Tidal effects would be less than those produced at the Earth's surface. An astronaut could pass through the event horizon without the tidal forces affecting him. He would not notice anything particular happen as he crossed the event horizon. The astronaut would have a few minutes to enjoy the experience of life inside a black hole before the tidal effects mounted to infinity."[9]

Like his floating astronaut, Roger reached a point where he could take a moment to survey his surroundings and appreciate the unexpected places his journey had taken him.

The concepts that bore his name—the Penrose triangle, the Penrose-Hawking singularity theorems, Penrose diagrams, the Penrose-Newman constants, and Penrose notation—had made him a prominent and successful "brand" within the world of mathematical physics. Roger pushed further into unexplored territory, letting twistor theory lead him closer to the fundamental geometric nature of reality—even as forces beyond his control threatened to tear his life apart.

In May 1972, the same month Roger's *Scientific American* feature appeared, Lionel died of a heart attack. In the months preceding his death, his mental acuity had declined significantly. He and Margaret had separated, and Lionel had become involved with a much younger woman who had worked for him as a research assistant.[10] Roger and his siblings had little to do with him.

It fell to Roger to sort through Lionel's estate. Among his papers, Roger discovered a recent letter that shattered him.

"My father was evidently talking about getting a divorce and this letter was from the solicitor discussing this issue," he said. Margaret knew about Lionel's affair, but escalating to a divorce would be devastating for her. "I retrieved the letter and put it in my pocket. Whether I burned it or whether I shovelled it in the middle of a lot of other letters, I haven't the faintest idea. But I didn't want that letter to be seen by her."

Roger, who wanted to end his own marriage, felt no empathy for his father. He recoiled at the idea that Lionel might find one last way to rob Margaret of autonomy. He identified with her, not him. He saw himself as trapped in a marriage with someone who wanted to control him and deny him a happy life. "It was like getting stuck in a spider's web," he recalled on many occasions.[11]

Joan and Roger could barely be in the same room together.[12] They slept and ate on different schedules, and Roger spent as much time as he could out of the house, at his office or walking in the woods. When they did cross paths, their interactions were devastating, not only for them but also for their sons. Aspects of their exchanges became seared in their sons' memories.

"'Stop your whining!' is exactly what he said when she was pleading for some emotional support," Toby Penrose recalled.[13]

Sometimes things descended even further.

"I witnessed Dad being physically aggressive to my Mum when I was a child. It had a very bad effect upon me," Toby said.[14]

"It was difficult to take, definitely. It was disturbing to see that sort of thing when I was kid. It kind of got into my system somehow," said Eric Penrose. "I did feel a sort of heavy cloud about the state of their relationship. It was becoming clear how bad it was from Dad's aggression."[15]

Roger offered few details about his treatment of Joan.[16]

I probably suppressed most of it because it was a time in my life that was very unpleasant. I do remember an occasion when I actually picked her up and threw her on the bed. There may have been occasions when there was no other choice. It's hard to explain because it was extremely difficult. I wouldn't say I was a saint. Certainly not. She was attacking me all the time—verbally I mean. As I say, it was like a spider's web. You can imagine an insect caught in a spider's web might well thrash around a bit.[17]

Whatever form Roger's "thrashing around" actually took, his and Joan's extreme unhappiness had a powerful effect on their children.

Toby felt emotionally rejected by both his parents.[18] "There was a singular blackness. An emptiness. A nothingness. In Dad's writing on black holes and in his verbal description to me of black holes when I was a child, I thought that it was perhaps a portrait of me that he was doing. I was the black hole that no light escaped from," he said.[19]

Roger was blind to his sons' trauma and Joan's pain. Even with decades of opportunity to reflect, the only regret he expressed was having married Joan in the first place.

He withdrew—into his work, his notebooks, and his conversations with Judith.

Her new job at the University College London Computing Centre made it possible for them to eat together frequently. Roger depended on those meetings to maintain his optimism and creativity. They often walked back to his office after lunch to so he could continue explaining things to her and describe how his book with Wolfgang Rindler was hobbling on.

A colleague at King's College, Malcolm MacCallum, agreed to turn his notes from one of Roger's classes into a major article on twistor theory. With Roger providing updates and inputs, their co-authored paper appeared in the prestigious peer-reviewed journal *Physics Reports*, which specialised in deep dives into major new scientific ideas.[20] They filled more than seventy pages explaining how twistor space could "quantize" classical theories of fields, space, and time.

Twistor mathematics appeared consistent with both general relativity and quantum mechanics. Like Isaac Newton uniting the heavens and Earth in a single gravitational theory or Albert Einstein revealing that space and time were part of a single continuum, Roger dreamed of using twistor theory to reconcile classical and quantum physics.

To solve twistors, he needed Judith. He was desperate for more access to her.

In August 1972, the NATO Advanced Study Institute invited Roger to deliver five lectures on relativistic symmetry groups at an eleven-day conference in Istanbul.

The conference abutted his and Joan's plans for their next North American trip, built around a visiting professorship Roger had arranged at Boston University. Joan and the boys would fly to Detroit while Roger was in Turkey. After the conference, he would spend a few days in Trieste, Italy, for informal meetings with colleagues. He would then fly from Rome to London and then on to join his family in Detroit.

Without Joan's knowledge, Judith and Roger made plans to meet in Trieste. Judith travelled separately from Roger and stayed at the same hotel in another room. His talks in Turkey went well, and he arrived in Trieste in high spirits.

He found plenty of time between meetings to go sightseeing and attend concerts with Judith. They wandered around the cliffside grounds of Miramare Castle and took a day trip to the port town of Muggia to read on the beach and talk about physics.

Judith had been dating someone in London for a few months. Roger didn't think it was serious. Here she was with him in secret at such a romantic locale. He saw it as a sign that she was at last ready to have a sexual relationship with him. On Saturday, their second-to-last night together, they escaped the summer heat, ending their day drinking wine in her hotel room. Roger believed the moment had come. He moved toward her and tried to kiss her.

She stopped him from going any further.

"This trip is just a holiday," she said.[21] She meant what she wrote in her letters: she did not want to have an affair with him.

As a physicist, he was attuned to the slightest whisper of the universe; the spin of an electron or the subtle bend of a light ray could shift his understanding of the world. But stubborn desire meant he could read her letters over and over and still not absorb this clearly delivered message.

He was angry and confused. This was not how the universe was supposed to be. He insisted on "some small token to enable me to continue with my own 'fantasy.'"[22]

He picked up a wine bottle as they argued and slammed it down on his own leg out of frustration. The bottle did not break, but the violence of the act frightened Judith. He left quickly, both of them feeling shattered.

The next morning, Sunday, he slipped a note under her door. "I feel rather rotten and won't bother with breakfast," he wrote her. "I would still very much like to sightsee with you today if you still want to do this. Call for me when you're ready.... I'll try and do some work or something in my room. (If you don't come for a while, I'll try you again.)"[23]

He wanted her to see how his actions reflected the power of his feelings and the importance of what he was asking of her.

I had always hoped—and still hope—that our problems are ones which can be cured by the passage of time. In fact, I remember feeling strongly, when I was in Canada last summer that you were the person whom (if nothing else) I would want to spend my old age with! Quite unfair on you of course, with me gaga at 72 and you a sprightly 59. But I'd rather not wait that long of course! And I know that it's not a practical thought. It wasn't meant to be. Just a romantic one . . .

I have loved you in some form or other for 22 years. I can't stop now.[24]

They managed a detente and spent their last full day together without argument. Judith rose early on Monday to see Roger off on the bus to Rome. Roger left unsettled and frustrated that he and Judith would have an ocean between them for the next several weeks.

Back in London, he was still in a dark mood. He reached Joan by telephone in Detroit. They argued over house logistics, the furthest thing from Roger's mind.

"This telegram came through over the phone [from Joan] the next morning. All about what to do with the mail, the neighbours, leaving lights on, curtains drawn (which way is drawn anyway? Closed or open?), take television back to rental people—oh yes and tell the police. I just felt myself fuming as the chap read it out to me.... I couldn't conceivably have time to deal with a list like that."[25]

This frustration partly reflected his turmoil with Judith. But he also was becoming increasingly swept up in the importance of his work, which diminished his patience for everyday demands. He felt he had infinitely bigger things to deal with than domestic tasks.

He spoke to his uncle Roland, who told him Margaret was upset about possibly not being able to afford to hold on to Thorington Hall. Despite his fondness for the place, Roger was unsentimental about losing it and thought it probably made sense to cede it completely to the National Trust. He assured Roland he would pay his share if the family decided to hold on to it.

Everything was a distraction. The only thing he really cared about was smoothing things out with Judith. He wrote to her on the plane to Michigan, attempting to focus on the best part of their Italian getaway: physics.

"I really think I understand Saturday night a bit better now, even after this short time in the bus and plane," he wrote. "My spirits were such that I could think again about the problem I started talking to you about while sunbathing in Muggia. That was the problem of how to do rest-mass using a series. Well, I think I solved the problem on the plane!"[26]

He devoted several pages to his progress on finding new solutions to the Dirac equation, which describes the behaviour of quantum particles like electrons and quarks consistently with relativity. Roger was trying to absorb his mentor's work into twistor theory.

He turned to more intractable problems. "Of course I know what's the big trouble with our relationship. I've known this for a long time. Perhaps I just attempt to ignore it when it suits me—or underplay it. But there's no easy solution anyway," he wrote.

The plane landed in Detroit before he could elaborate. Roger and Joan still had their old Saab stored at Joan's parents' house, and after one night at his in-laws, Roger drove Joan and the boys to Blink Bonnie. The Canadian leg of the trip was a holiday for everyone except Roger, who planned to write up his Istanbul talks and mail drafts to Judith for her reaction.

Joan was happier for being away from London, but Roger was still in a deep funk.[27] Trieste hadn't gone the way he expected, *Spinors and*

Space-Time felt no closer to completion, Detroit was miserably hot, and Canada was too expensive.

He needed to resolve things with Judith. He could no longer imagine carrying on his work without her.

Late one August night, with Joan, Christopher, Toby, and Eric asleep, he began a letter that he told Judith, "I *must* write."[28] He knew things had gone very wrong in Trieste but believed there had to be a solution. He had solved black holes. Could his and Judith's problems really be more difficult than that?

He approached their situation methodically, eliminating variables and weighing solutions. "There might be a choice, either to decide never to see you again, or to decide to pretend and to keep a relationship going on some low-key but false basis, or to be bold and try to dig deeper into the foundation of our relationship," he wrote. Parting ways would be too painful, and sublimating their incompatible expectations was impractical. The only way forward, he concluded, was to dig deeper. Anything else would be "throwing overboard a fantastic opportunity."[29]

He was unhappy that she didn't want a physical relationship but outright panicked at the thought of losing his source of scientific inspiration.

> On occasion I have had the feeling with you that "together we can conquer the Universe." It is not even just my work on twistors that I am referring to.... It is something rather vaguer and perhaps larger that I mean. I'm not even sure what it is, but I have this small hope of being able to help the world along in some small way. But first one must understand. And with you I can gain understanding of things in a deep way. It is your unique combination of human warmth, personal understanding, and appreciation of the artistic values in mathematics which is so valuable to me.[30]

He couldn't understand how their joyous, productive relationship could have derailed so thoroughly.

He acknowledged their age difference, their complicated history, and "very real other problems." "I *am* married. And I have three children who are still young. And the kind of life that you want to lead may not fit in with mine. And there *is* a kind of 'responsibility' involved, I suppose, in getting entangled with me."

He said the most important problems, though, were Judith's hang-ups. "I hope you will forgive me if I concentrate on what I see to be some of your own internal demons," he wrote. "I may get them wrong. I am sure that there is some important truth in what I say, but equally sure that this truth can be only partial. So take it only as food for thought."

He had long known, "but selfishly chosen to ignore," the impact of having been a "father substitute" when she was a child. He acknowledged it now so he could ask her to get over it. "I think now that I might possibly be able to continue on almost any basis. Pain there could be, but this has to be set against the bleak vacuum and lost chances of the alternative. If our meetings cease, a spark of life goes too."[31]

At 4:00 a.m., Roger sealed up the envelope, slipped out past his sleeping family into the cool night air, and padded down the road in pyjamas and bare feet to pop the letter in the post box.[32] He was finally ready for sleep.

Four days later, he wrote again. "Enclosed, please find one (1) deeply personal document for your perusal. It's nothing to be frightened of."

It was the write-up of his first Istanbul talk. As he had done with *Spinors and Space-Time*, he characterized this work as a letter to Judith—an encapsulation of the emotional and intellectual bond they secretly shared.

"The theory of twistors is a formalism for relativistic physics which affords a new approach to the description of quantized fields and space-time itself. In the twistor formalism, space-time points need not be employed as the primary objects in terms of which all else is to be expressed. Instead, the primary objects can be the twistors themselves," he wrote.[33]

Roger and Judith were the only people who could see the love letter hidden in his treatise on advanced theoretical physics.

Each time he wrote to Judith, Roger's spirits lifted. It seemed to him not only that they *could* resolve their differences but that they already had. The mere thought of her reactions led him to new ideas and approaches—even before she wrote him back. "Your magic still seems to work with me even though your response...is (as yet) only imagined."[34]

He sent her pages filled with new equations, diagrams, and half-formed thoughts on twistors and related ideas. He repeatedly assured her that, while he welcomed her reply, she needn't respond in detail. "I am only asking myself these questions 'out loud' so to speak."[35]

As his mood lifted, Roger also got along better with Joan and engaged more with his sons.

The boys discovered a fallen squirrels' nest beneath a tree near the cabin. Most of the kits had died, but one baby squirrel survived the fall. They adopted it, organizing with their parents to nurse it back to health. They found a box and bedding to keep it warm and safe. Joan, Christopher, Toby, and Eric cared for it during the day. Roger tended it at night during breaks from his writing. The family bonded over the effort to save this tiny animal.

Their holiday wound down, and they prepared to travel on to Boston. On their final night in Bayfield, Joan and Roger hired a babysitter and drove to Stratford, Ontario, to see a performance of Shakespeare's *As You Like It*. Keeping Judith's presence in Trieste a secret was eating away at him. Possibly influenced by the play's convoluted stories of infatuation, love, and marriage, Roger resolved to come clean with Joan before they left Ontario.

He confessed on the drive back from the play. Judith had been in Trieste. Nothing sexual had happened between them. And he might not meet with her anymore. End of story.

Joan had more questions than accusations, which Roger found a tremendous relief.

They entered Blink Bonnie with Joan still working to understand exactly what had transpired between Roger and Judith. The sitter greeted them with a grave face. The baby squirrel had died while they were at the play.

"I cursed my lousy timing....Joan was certainly very upset about the squirrel. But she also questioned me closely about Trieste," he wrote.

Roger told Judith he believed Joan was more bothered about the squirrel than Trieste. He also thought his wife was "genuinely upset" at the prospect he might no longer be able to meet with Judith.[36]

Nothing about Joan's behaviour backs up his impression. She had been worried about Roger falling in love with Judith since they reconnected. She may have been too resigned, frightened, or insufficiently certain to push the matter that night, but she believed they were having an affair.

Still, Roger left Ontario believing he'd assuaged her suspicions.

In Boston, the university provided them with a two-bedroom suite at the Brookline Motor Hotel on Beacon Street, a major thoroughfare near the campus. The stuffy mini-apartment had too little space for five people. The door to the children's bedroom didn't close properly, the stove and sink jutted into the hallway that led to the only bathroom, and there was nowhere for Roger to retreat. He only found peace and quiet late at night when everyone else was asleep. Joan sometimes slept in their bedroom, sometimes on the couch in the small living room. Whichever room she left empty, Roger used to work.[37]

He continued writing to Judith every few days, offering new ideas and repeated thanks for the feedback he was anticipating. His tone started to change, though, as he continued not to hear back from her. "I won't be swamping you with 'research thoughts' this time," he told her. "I'm not sure how well I can keep that lark going until I get some verification of your continued existence!"[38]

He also exhorted her to hold on to his letters and drafts. "Someday I may want to look them over again to search for ideas, so please keep them."[39]

As mid-September passed and her unresponsiveness persisted, he grew increasingly despondent. He started to question what he was even doing in Boston, given "the lack of any real significance" in his talks and conversations, an ever-growing antagonism toward America in general, and the claustrophobia he felt crammed into this tiny apartment with his family.[40]

Consumed with a "torment from the soul," he couldn't stand the silence any longer.[41] He sent a telegram to Judith (falsely) saying his original copies of the Istanbul lectures had been lost and he needed her copies back. He was certain this would force a response. Judith was out of town when the telegram arrived. Her roommate Gill received the message. Gill telephoned Shirley, who was supposed to be meeting Judith at Thorington Hall for the weekend, but Shirley had cancelled at the last minute. Days passed before the message got through to Judith, by which time she felt there was no point in replying—if Roger were that worried about her or the papers, he would no doubt phone to follow up.

While Roger waited for his telegram to have the desired result, a letter Judith had written weeks earlier arrived at his office at the university. She had mailed it to Bayfield, where it had just missed him. After several days of delay, it had at last made its way to Boston. Two more letters arrived the next day.

His calm and optimism returned. He was so delighted to hear from her, it took him several readings before he noticed she was still angry with him.[42] She rejected his dismissal of their history and his marriage as mere "practicalities." And she cared even less for his analysis of her "demons" and "fantasies."[43]

"Only two things I'll say now—immediate reactions. Firstly I am very glad that you reject the 'fantasy' solution to the problem," she wrote. "We are both too honest to have...a very superficial relationship based on preserving each other's fantasies. Secondly...I disagree (as you thought I would) with your analysis of my feelings, or most of it anyway."

He apologized for his "two-bit psychiatry" but expressed no regret about his earlier letters, given how important she was to his work.[44] He was embarrassed his telegram had revealed what a "worry-nut" he could be. But he carried on psychoanalyzing away at Judith's resistance to their relationship.

He both expanded on his love for her—"It is real love; deeper adult love than I have felt for anyone before—or expect ever to again. It is even love...of a kind that I knew not that I had in me"—and also sent her copies of the next two Istanbul lectures, exhorting her to continue motivating his work.

"I have often, in the past, tried to think of circumstances which produce good ideas. How can one 'maximize' the potential for ideas? Everything I tried failed. Good ideas just seemed to come at random—except, of course, that one must be thinking about the problem. But with you it has been different. For the first time, a deliberate action on my part (i.e., involving you in my research) has produced genuine inspiration."[45]

Judith entertained Roger's scientific needs but held firm on everything else. "I like you 'airing your ideas' to me, as long as you don't need a reply (though you may yet get one). I hope it does really help you to write to me—if it was a disguised effort to get letters back, I would feel very guilty," she wrote.

She reminded Roger that she too was in a relationship. "In all your writings and talkings about me, and us, you never mention Colin," she said. "Do you feel he's got no bearing on anything, or that he doesn't matter, or is it just part of your fantasy that he doesn't exist? If we are going to delve deeper, we must both realize that he is very relevant."

Roger didn't think Colin was relevant. He couldn't take seriously such attachments to the mundane world. True, she had the demands of graduate studies and her job. True, she and Gill had a steady stream of friends they went out drinking and dancing with until late. True, she worried about money, health, friends, and shopping. But he couldn't see how any of that really mattered.

On rare quiet evenings or when she took a sick day, she found time to read his manuscripts. As his work became more ambitious and advanced, though, she struggled even more to understand what he was driving at. "I'm afraid I have to confess that I've been unable to follow any of your twistor thoughts in your letters. Either I've forgotten it all, or it's much easier when you're there and I can ask you things," she confided.

At his request, she had photocopied some of his notes for another London-based researcher to read. "You're handwriting...is always unmistakably yours. I've had to trace parts of it where the copying hasn't worked—it's a strange feeling. As you know, I like your writing and to trace it made me feel very close to you."[46]

Guilt and insecurity filtered through her letters. She packaged her edits with a cushion of apologies—"don't take anything I say too seriously"—before pushing him toward greater clarity and brevity. "What you say on page 1 about local isomorphisms isn't at all clear.... When you lecture, did you really go through the material of pages 12–18 with no words or pictures or examples of usage?...You keep saying, 'I shall explain shortly/later—it's a bit annoying to read.'"[47]

Just as gingerly, she worked to sort out their relationship, trying to be firm while assuaging his feelings. "Like anyone I'm interested and flattered when the subject under discussion is me. Even if I totally disagree with you, it's good to hear your point of view because, in my rejection and argument, I learn something of myself, and from your opinion I learn something of you."[48]

She apologized for sending mixed signals, for being moody, for overreacting, for changing her mind about "acceptable compromises." Still, she held steady.

Roger, I hope what I'm about to say isn't too painful—I hope if it is you can temporarily reject it and absorb it slowly, later.... What Trieste showed me was that there is something I need which you haven't got. There is the one aspect, which I think you were saying in your letter, that it's a bit "shocking" for me to think about you, who has been a sort of father and brother to me, in a sexual way. I am frightened of any sexual contact with you. Can I ask you to accept that, to bear with my neuroses?[49]

She wasn't saying anything new. But pressure from Roger filled her with guilt for once again needing to say no to him. The last of the flurry of letters he received in Boston ended with her saying, "I'm not sure how but I feel I've behaved very badly towards you. Please forgive me."[50]

Roger sank deeper into his work. Each Istanbul write-up he sent was longer and more in-depth than the last. She understood less and less with each instalment and found his writing no longer approachable or engaging. She worried Roger had lost his touch—or she had.

"What was wrong? It sounds awfully conceited but particularly after all you said about my influence, I can't help worrying that that awful last night in Trieste could have been responsible. Really it reads rather like a section in a maths book which is getting some rather dull calculations and formulae out of the way before it can get on to the really interesting stuff, if you see what I mean. Where is the spirit, the freshness that has always marked your seminars?"[51]

In fact, Roger showed no sign their argument in Trieste had done anything but encourage him to be more ambitious and more confident in his work. She thought he was brilliant and his ideas important. Surely his colleagues would see the same thing, no matter how dense his writing became.

He was equally optimistic about his personal life, despite having made no progress in either improving or ending his marriage.

Judith encouraged him to sort out his relationship with Joan independently of his feelings for her. He wasn't sure that made sense. In a deterministic universe, could he really take ownership of his unhappy marriage? The idea felt strange to him.

"[Marrying Joan] was a 'deliberate' action on my part which (and I must qualify this statement in a moment) to the best of my present judgement, I cannot help viewing as a 'mistake.' Unless I take the sort of deterministic view of life which removes from me all responsibility for all actions (and sometimes I find myself doing just this), I must view whatever consequences I see of these actions as, somehow, 'my responsibility.'"

He recognized he was "trying to take refuge in impersonal philosophic generalizations" but resisted the idea that his marriage was *a* mistake, let alone *his* mistake.

He examined the counterfactuals. Would he have ended up with Judith if he hadn't married Joan? Would he have been happier? He and Joan had produced "three wonderful, intensely individual delightful (usually!) talented, intelligent, and highly original little human beings."[52] But would he really miss his sons in an alternate timeline in which they never existed? His children mattered to his *actual* future but not to other hypotheticals. In reflecting on these unanswerable questions, he hoped it wasn't all his fault.

"It does something good for my ego if I feel that my action was not viewed necessarily as idiocy from the outside," he concluded.

This insecurity sometimes spilled over into his science work. Even with Judith's encouragement, he doubted his purpose, especially when Joan was impatient with his detachment from everyday life. "This has made me feel guilty about my own mode of thought and of working, which has always been terribly 'inefficient.' I can sit dreaming for hours about matters quite irrelevant, feeling guilty all the time for not doing those things which most immediately need to be done," he wrote.

He also worried about twistor theory becoming *too* successful. If other mathematicians and physicists took up the cause, he might lose control of his most prized creation.

> I have always worked very much on my own, with my own peculiar and individualistic points of view. This applies particularly to the twistor theory....For a long time, the interest of others in this subject has been slight. Although this entails a feeling of being "not appreciated," there is, nevertheless, something very comfortable about that state of affairs. I can work at my own pace, doing what interests me, and not having to worry about competition

from outside (particularly, about competition from the "research factories"). I do not really like competition. I never have. I tend to stay away from hotly contested fields. I have recently heard word of some of the most unlikely people starting work on twistor theory (e.g., Bryce DeWitt—of "quantizing-general-relativity" fame), and even worse, setting their research students to work on it.[53]

Roger had watched Stephen Hawking take the Penrose singularity theorem in new directions to public and academic acclaim. The issue of the surface area of merging black holes still rankled him as well. It troubled him how his best ideas sometimes slipped away from his control.

Ultimately, Roger's professional restlessness provided him the exit from his marriage.

In early October, Judith sent news that Fred Hoyle was retiring from Cambridge, vacating the prestigious Plumian Chair of Astronomy and Experimental Philosophy. The position appealed to Roger much more than anything available in the United States.

"Perhaps I regard this as my 'salvation,' not only as an antidote to having to join a research factory in the USA—but also (and here's some honesty for you) because of Joan's expressed dislike of the Cambridge atmosphere and the hope that she might leave me if I went to work there."[54]

If Joan left him, he would no longer have to ruminate over whether *he* needed to take action. If fate took him to a city she didn't want to live in, their relationship could have a natural ending that placed no burden of responsibility on him.

Two weeks later, Roger heard from Dennis Sciama that a second highly prestigious position had become available: the Rouse Ball Chair of Mathematics at Oxford, awarded to "a mathematician who applies modern mathematical ideas and techniques to the study of one or more aspects of the physical sciences." Sciama said it was tailor made for Roger. Roger couldn't help but agree. It was a much better fit than the Plumian chair, though it was located in a city that Joan had no particular aversion to.

The fates decided—Cambridge offered the Plumian chair to Roger's old friend astrophysicist Martin Rees. Roger was offered the Oxford position.

Joan and Roger returned to London in November. Joan began seeing a therapist. At Joan's request, Roger joined one of their sessions, at which they confronted Roger, asking him to admit to having had an affair with Judith. He denied it.

"Don't lie to me," Joan implored him.

"I'm not lying. This is the truth," he said.

In his mind, the fact he had not had sex with Judith meant there had been no affair—despite his pining for her, meeting with her in secret, and sharing such intimacy.

"She wouldn't believe me," he said. "Eventually, I walked out."[55]

Joan found a second therapist, who offered to speak to Roger one on one. Still smarting from his previous experience, Roger readied his defences, entering the session "armed with all my rationalizations and justifications, imagining that I would be hard pressed to get him to see anything of my own perverted point of view."[56]

But to his surprise, the counsellor—a man this time—seemed to recognize the marriage was "basically dead" and, even better, didn't treat it as Roger's fault.[57]

"He just said, 'Look, you've got this job at Oxford. You take this opportunity as a way of getting out of this relationship.' That's what he told me," Roger later recalled.[58]

The therapist gave him the external permission he sought to separate from Joan at last. The universe had spoken. He moved to Oxford on his own.

13

APERIODIC

The world had closed in tightly around Roger for so many years, he had almost forgotten any other existence. Hiding from his family under the living room, cramming onto the Tube, pushing papers around his office to find a clear spot to work. Life now seemed to open out in front of him, offering space, silence, and more time to focus.

He hadn't told Joan he was leaving her. For several months, Joan remained under the impression she and the children would be following him to Oxford. She, along with Christopher, Toby, and Eric, believed there was still hope for the family to stay together.

"I think he wanted to get away from her quite badly and Mum didn't want it to end," said Toby. "I wanted them to be together. They seemed to get on when they talked about politics and other topics, though the atmosphere between them was frequently tense and oppressive."[1]

Roger appeared to have completely absorbed his own father's inability to state matters directly, leaving his family to realize on their own he didn't plan to live with them again.

Margaret also found new freedom. Max Newman's wife had died about a year before Lionel. He and Margaret reconnected at Lionel's memorial.

"I remember, he'd had his eye on Margaret from way, way back. There was some indication of that. Margaret had been taking to the bottle, somewhat, drinking a lot of whiskey after Lionel died. But then Max completely replaced the bottle," Roger said.

Less than a year after the funeral, Margaret sat with Roger and asked out loud, "Should I marry Max or not?"

"She went through this list of complaints—'Oh, he's such a fusspot!' She got to the end of the list and said, 'I think I will!'"[2]

Margaret and Max married when she was seventy-three. No longer a fly on the wall, she felt fully human once again, engaged with the art, music, maths, and games she had always loved.

"Margaret could express herself again. She started playing the cello, which was something that would never have happened if Lionel had been around," Roger said.[3]

Roger used his new freedom to dive deeper into work. After a decade of development, twistor theory had yet to find traction beyond a relatively small number of pure mathematicians and physicists. He still worried he could be wasting time on "some amusing side issue" that might never prove useful in the real world.[4] Even in these moments of doubt, though, twistors continued to yield promising new clues and point him in new directions.

Judith had moved out of London to a flat in Haywards Heath, commuting nearly sixty kilometres every weekday to her job at University College London.

As Joan came to understand that she and Roger were separated, she occasionally vacated the house on weekends, allowing him to spend time with his sons. If he wanted to see Judith, it meant an extra night at Stanmore. He preferred staying over on Sundays rather than Fridays, so he could watch television with the boys. "The call of Monty Python is strong!"[5]

Between her job, studies, and friends, Judith's time for Roger was limited. When he couldn't see her in person, Roger wrote letters.

When he wrote for publication, he still pretended he was writing to her.

"It's all complicated, and it's all connected," Roger often said.

From the spin of a single electron to the distant collapse of an eleven-billion-year-old galaxy, Roger drew on every source he could find to graft twistor theory onto the entire length, breadth, and depth of the universe.

So focused was he on twistors, he barely noticed a new idea worming its way into the back of his mind—a seemingly fanciful side trip that became one of his most unexpected and influential creations.

Roger found these side trips essential. He stored a lifetime of half-formed ideas that periodically bubbled up to the surface. Sometimes they produced only unproductive distraction. But other times, an unexpected catalyst turned one of these ideas into something amazing.

In 1973, while Roger was transitioning to Oxford, he received an invitation to speak at London South Bank University. The content of the invitation was unremarkable—Roger received dozens like it from across England and around the world. He had ready-to-go talks

The coat of arms for London South Bank University.

on black holes, twistors, and other favourite subjects, with stacks of transparencies that could be combined and recombined at a moment's notice.

He tossed this invitation on the pile but paused periodically to examine the university logo on the front of the notecard. Its red and gold shield featured six pentagons forming an angular flower with narrow triangular gaps between each petal. This pattern was well known to Roger—he had traced it out many times making polyhedra with his father. Folded in three dimensions, the six pentagons formed half a dodecahedron.

He noted how the pentagons and gaps formed most of a larger pentagon. He pictured what would happen if he subdivided each petal into smaller flowers. Or placed six of the large pentagons together to make an even larger one. Flowers within flowers within flowers. An unconscious shadow of a memory from Lionel's manuscript by Johannes Kepler hovered in the back of his mind.

"I had seen Kepler's images, but I wasn't thinking about them when I produced my pictures. However, I would consider they had a psychological influence on me to the effect that pentagons were not a dead loss," he said.[6]

Weeks passed, and he kept the invitation handy, revisiting it in idle moments, sketching pentagon patterns in his notebook, even when he knew he should be doing something else. "What I *should* be doing and what I *was* doing were not very clearly divided," he recalled.[7]

He felt as if parts of his brain that had been devoted to misery were now redeployed to think about nested pentagons.

Judith noticed the difference, writing, "Perhaps you have been a little less demanding, less intense but I think it's more that I am learning to accept and enjoy our relationship. The double guilt I felt about hurting Joan and hurting you seems to have been submerged, perhaps just because it's gone on this long and no-one seems to have been hurt. Perhaps because now I realise that I need to go on seeing you, I can rationalize the guilt away."[8]

Roger also felt less pressure to keep his relationship with Judith a secret.

The summer before he began lecturing at Oxford, he travelled to Austria for several weeks with Wolfgang Rindler. They resolved once again to finally finish writing *Spinors and Space-Time*. Wolfgang's parents lived in Vienna and travelled to their summer home in Baden on the weekends. Roger and Wolfgang did the opposite, working undisturbed in whichever residence his parents had left empty.

Early in their stay, Roger shyly disclosed to Wolfgang the existence of Judith. It felt freeing to at last be able to talk about her.

"Telling Wolfgang something of you makes me less inhibited about propping one of your pictures up in front of me when I write—although I still only dare do that in the small hours when he's asleep! But it makes the night hours and the writing more enjoyable for me."[9]

He and Wolfgang fell into quiet, simple routines.

There's something comfortable, in a way, about having such a well-defined goal which allows me to put out of my mind all the worrying thoughts about life. Even just the process of living is very easy. Wolfgang does virtually all the chores—and there's not really much of that. Just washing a few dishes or making a pot of coffee and the like. We tend to eat at most one meal out, living off bread, ham, and cheese at other times. I wouldn't like to live like this all the time! But as I say, there's something relaxing about it for a short while.[10]

He felt carefree and content to take his writing wherever it led him—until Wolfgang gently led him back. "Roger is at heart a mathematician who loves playing with things," Wolfgang recalled. "He often went off on beautiful tangents, which didn't really interest me very much because they weren't physical."

In July, the book still unfinished, Roger prepared to fly back to England. Judith was eager for his return.

"I feel the need to write because I miss talking to you so much. Only a week today and I'll be seeing you for lunch! I don't know why but I've felt the lack of our meeting far more this time you've been away than when you've been away for far longer in the States. I think I was mentally unprepared for it, and of course moving to Haywards Heath has created a lot of things to tell you," she wrote.[11]

Judith felt she and Roger had reached a resolution around their friendship, but the rest of her life remained demanding and difficult. With work, studies, and learning to drive, she often didn't get to bed until 2:00 a.m., rising again at 6:30 to catch the train into London.

She arranged to meet Roger at his London office the first Monday he was back but warned him that she was having a difficult week.

> Tuesday was the worst. I don't usually mind the heat, but it was too hot even for me, especially as I was rather tired and my period arrived, so I wasn't even facing the day with any bounce. I worked very hard all morning, spent the lunch hour trekking round shops getting material and things for the flat, and then I'd done another couple of hours hard work when I discovered that, due entirely to a stupid error on my part, I had deleted everything I'd done that day. Feeling near to silly (menstrual caused) tears I thought, "Sod it, I'll stop and have a cup of tea." There was no tea but, undeterred, I decided to have coffee. Somehow, I still don't really know how, I scalded my arm on the steam from the kettle.[12]

The burn needed to be dressed at the local hospital, and the work she had lost at the computing centre needed to be redone. Roger's return couldn't come soon enough.

"I hope it comes through between the lines how very much I miss you and what and important part of my life you are. I'm longing for Monday. You give meaning to it all. My life is O.K. but pretty superficial without you. Thank you for your letters, for being you, and most of all for loving me—I can still feel wonder and awe that you should care about me."[13]

By October, Roger had found a one-bedroom flat in a house on Winchester Road, a quiet five-minute walk from the Oxford Mathematical

Institute. His briefcase quickly became overstuffed with "a depressing pile of urgent matter still left undone."[14] But the demands of his new position weighed him down far less than life in London had.

The tools of his research remained the same: blackboard and chalk, a journal always within arm's reach, conversations with colleagues, and his letters and meetings with Judith. His mind felt sharper and more active.

He no longer had the original invitation from London South Bank, but his journal was now filled with pentagons, arranged in repeating flower patterns, with gaps forming triangles, rhombuses, stars, and jagged, asymmetrical polygons. Unlike the hexagon tiles on the backsplash in his house in Canada or the interlocking animals in an M. C. Escher print, pentagons didn't settle down into a repeating pattern—they didn't tessellate.

He discovered six essential tiles: three types of pentagons differentiated only by the shapes they could abut and three shapes formed by combining the gaps between them—a five-pointed star, a rhombus, and a partial star he called a *jester's cap*.

Following very simple rules, these six tiles created interlocking decagons, pentagonal flowers, starbursts, and twisting chains that expanded across the plane, creating an infinitely non-repeating pattern with no gaps or overlapping tiles.

The result was hypnotic. Penrose tiles invited a search for repetition that did not exist.

Tiling games predated Roger by millennia. Ancient Greek geometers covered surfaces in elaborate tessellating patterns of squares, rectangles, rhombuses, and triangles. Fifteenth-century Islamic architecture featured "Girih tiles," whose five-axis symmetries were closely related to Roger's creations.

In the twentieth century, tangrams, polyominoes, and other tile-based puzzles and games gained popularity among recreational mathematicians.

In 1961, logician Hao Wang created a set of multicoloured square tiles. His *prototiles* were all the same shape and size, differing only in the colours of their sides. He stipulated that the tiles could not be flipped or rotated and must be placed so that the colour on each side matched that of its neighbour. He conjectured that any given set of Wang tiles that could cover the infinite plane must have a "periodic" pattern: it would reach a point where either no further tiles could be added or the colours would start repeating.

Testing the Wang conjecture required forays into formal logic, computer science, and philosophy.

One of Wang's students, Robert Berger, disproved the conjecture using the "deterministic automata called Turing machines."[15] It was impossible for a computer algorithm to determine whether each subset of Wang tiles was "solvable" for the infinite plane—another example of the Gödel-esque incompleteness of mathematical systems that had fascinated Roger since graduate school. Berger also concluded there must be a set of *aperiodic* Wang tiles that could *only* cover the plane non-periodically.

Berger found a set of 20,426 aperiodic Wang prototiles. He subsequently presented a dramatically simplified set of 104 aperiodic prototiles, a massive achievement that inspired other mathematicians to compete in earnest to find the smallest possible aperiodic prototile set.

Using pentagon-based tiles rather than squares, Roger got it down to six.

He needed to tell Judith, but there were new complications in their relationship. She worried he didn't have enough of a life beyond her.

"It frightens me to feel that, apart from the children, you feel what you feel almost exclusively for me, whereas I feel, at many different levels and in different ways, for several people—I don't just mean Colin, I don't even mean Colin!—but there are people as important to me as you are. And perhaps one source of my odd feelings of resentment towards you is the feeling that I'm neglecting them because of you. Without really knowing what I mean can I ask you once again to ease up a bit?"[16]

Roger didn't know how to ease up and, in trying to explain why, revealed that he still held out hope for something more from their relationship.

The trouble was (and still is, in a way) that I cannot always tell what actions on my part constitute pressure on you. Although to be with you is what I want most...there is something of the pride in you that I feel, which makes me want to show you off to the world. And there's another thing. I would like somehow to tell the world of the inspirational value that you have had for my work—even to shout it from the mountain tops—and to increase that inspiration and to shout that even louder from the mountain tops. There is a frustration in the fact that I have had to keep it secret.[17]

Judith had exams coming up. Roger agreed to a "moratorium" on any mountaintop shouting until she had made it through.

"I have a mark down in my diary against 14th June. Is that the date of your exam? I would love to book you for a dinner that evening. This would not be to discuss any terrifying matters, but just to celebrate and to enjoy your company in a relaxed fashion!"[18]

He pressured her for more than time. He wanted to marry her and start a new family.

"You know that I have always wanted a daughter," he wrote. "And I feel I want to mould ourselves into one composite person. Perhaps that is what marriage and having children is emotionally all about!"[19]

At a lunch on March 18, 1974, she girded herself for a painful conversation. She reminded Roger how she and her first husband had been unable to conceive. She had been having irregular, painful periods and terrible mood swings. Her doctor had told her that morning she required an operation (likely a hysterectomy) that would mean she would never have children.

Roger went quiet, unsure what to say. He wrote her later that day.

"O Judith, I do feel for you so! I know that having children of your own was something you had seen as a wonderful hope for the future. And it was a hope that I could share in," he wrote.

He could not resist reflecting on his own loss.

I am not able to tell how this news affects our relationship. I can see factors on both sides. So my reaction to the news must be independent of all that. Except, I suppose, for the small fantasy hope that I might someday have a daughter by you....Judith, I just wanted to say one final thing, in case it helps you to view the future. I am not asking you to say anything or commit yourself in any way whatsoever, but I wanted to let you know that the news you received at the hospital does not in any way affect my hopes and desires about the possibility (or possibilities) of sharing a life with you. I still mean it as much as ever. My offer to marry you as soon as I am able still stands. But I am only saying this to you for your own information. I am not asking you for an answer.[20]

Four days later she wrote back.

On Monday and Tuesday I did go through some pretty doom-laden patches—feeling very bitter about the futility of all the trouble, physical and emotional, I've had for the last 15+ years with that part of my anatomy. And also feeling unable to visualize a future at all. I can't say it's shattered all my plans, but it's removed the central theme of most of my daydreams. I just can't accept myself as a "career woman." It's not that I feel "no one would love me" in my present state but, for myself, I can't see any reason at all now to plan a future in terms of marriage.[21]

She busied herself with work and studies, got drunk, visited her sister's family. Her sadness was overpowering. "I'm still physically not really well and at college I found it an effort not to scream or cry and I wandered about hunched up over my offending tummy, not speaking, not smiling—a right old misery!"[22]

Roger's attempts at support gravitated back to his work. "I want to say that if it is your fate that you cannot have children of your own, then I hope that you can feel as I do that something I am trying to offer you is a possibly acceptable substitute. I don't know if I can explain what the 'offer' really is. . . . Perhaps I can best describe it as a romantic adventure. It is, among other things, an adventure into the world of unknown ideas."[23]

He was worried about her and also about *Spinors and Space-Time* stalling out again.

"Things just haven't been going as well as before," he wrote a few weeks later. "I find myself just looking forward to the time when circumstances force us to stop, rather than looking forward to having everything finished. I don't think it's just my mood. . . . It's more that the work is just intrinsically more boring and tedious."

He let her know that Shirley had had a baby girl on July 9 and that he'd made progress with his pentagons.

It had started with a serendipitous conversation with a Belgian-born, Princeton-based mathematician named Simon Kochen, who was spending a year as a research fellow at the Oxford Mathematical Institute. Visiting researchers worked two per office, which Simon found intolerable. "I can't work with another person there. I ended up talking too much with them," he said.[24]

Roger and Simon had met years earlier at Cornell. They shared an interest in integrating mathematics into physics, philosophy, patterns, and play. Roger was rarely in his own office and offered it to Simon as a quieter base of operations. Whenever Roger stopped by, Simon couldn't resist putting down his work to chat.

Simon gave Roger a recent journal paper that described another set of six aperiodic prototiles, tying Roger's feat.

Roger skimmed the article and told Simon, "I can get it down to five." The jester's cap and one type of pentagon always joined each other the same way. He could fuse these two shapes together, reducing his prototiles by one. He hadn't done it before only because he thought the combined shape was ugly.

But was five really the lower limit?

Late into the nights that followed, his thoughts wandered back and forth between Judith and his tiles. Miraculously, he could feel her magic unlocking doors.

He posted a letter to her hospital room, where she was recovering from surgery.

> Sometime in the middle of the night (morning) after I saw you on Friday when I tried wake myself up to do some work on the book, I started thinking about that pentagon pattern again. (My mistake was to show you those proof sheets!) I knew that by sticking some of the pieces together I could reduce the number to 5. I suspected I could get it down to 4. Well, in the middle of the night I satisfied myself (pretty well) that I could do it in 4 pieces. But then I realized that more simplification and a little chopping and changing enabled me to get it down to 2 pieces![25]

He sketched out the two new prototiles. They looked nothing like the previous set—less jagged and simpler. Each had only four sides. The five-axis symmetry that gave them their aperiodicity was discreetly encoded in their lines and angles. Roger used dots and shading to establish how they could be legally placed together—gray had to match with gray, speckles with speckles.

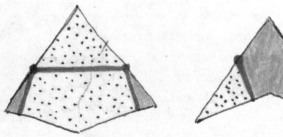

The original sketch of kite and dart tiles, from Roger Penrose's 1974 letter to Judith Daniels.

"The remarkable thing is that they automatically assemble the pentagons, stars, diamonds, and half-stars of the previous pattern," he wrote. Judith was the first person to see what would become known as the kite and dart tiles, arguably Roger's most famous creation.

Two simple shapes cracked and divided the infinite plane, exploding ever outwards in unending variety. No tile set could approach their extraordinary, unexpected simplicity.

Johannes Kepler, London South Bank University, and Simon Kochen had all contributed to the Penrose tiles. Roger, though, treated Judith as his primary source.

"Your influence...enables me to solve problems even when I'm not trying to!"

In late 1974, Roger gave a speech at a conference organized by the Institute for Mathematics and Its Applications in which he came close to publicly acknowledging Judith.

"Recently I wanted to design something interesting for someone who was in hospital to look at and I realized that there was a certain definite rule whereby one could continue such a pattern to arbitrary size," he told the audience, consciously omitting the sex of his hospitalized friend.

The talk, "The Role of Aesthetics in Pure and Applied Mathematical Research," was a sort of manifesto in which Roger resolved issues he had grappled with all through his mathematical career: What makes a problem worth pursuing? Does it matter if a question is merely interesting rather than useful?

> I think when we talk about the justification for doing mathematical research, first we have to ask what is the justification for doing mathematics at all?
>
> Often in mathematics, one talks about whether a result is useful or not, meaning whether it is useful to other areas of mathematics. To a working mathematician this seems worthwhile in itself and he does not feel that he has to justify it especially. But

I think...the man in the street might feel that the justification lay solely in being able to do bigger and better calculations, so that stronger and cheaper bridges or aeroplanes etc., could be constructed. Of course this is part of the story, but for a mathematician it is only a rather small part....Basically, the motivations turn out often to be ultimately aesthetic ones; so often a subject is pursued simply for the pleasure that it gives. Perhaps this is a pleasure that one feels because one is searching for truth.

I think that if the subject is to be judged solely on this basis, one would have to compare it, perhaps, with some very off-beat, avant-garde form of art where the people who could really understand what was going on were very few indeed....But set against the matter of the very small audience, mathematics does have the advantage that directly or indirectly, and often in unexpected ways, it has applications in many other fields.

I have been concerned with aesthetics in its role as a justification for what one is doing. There is also a more subtle role played by aesthetics in connection with research, namely as a means for obtaining results. This is really an important aspect of the role of aesthetics which I do not think has been discussed very much. It is a mysterious thing in fact how something which looks attractive may have a better chance of being true than something which looks ugly.

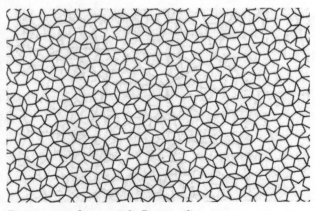

Finite section of an aperiodic Penrose tiling.

I have noticed on many occasions in my own work where there might, for example, be two guesses that could be made as to the solution of a problem and in the first case, I would think how nice it would be if it were true; whereas in the second case I would not care very much about the result even if it were true. So often, in fact, it turns out that the more attractive possibility is the true one.... Most guesses are likely to be wrong anyway. But the proportion of correct guesses among the attractive possibilities seems to be far higher.

. . .

Elegance and simplicity are certainly things that go very much together. But nevertheless it cannot be quite the whole story. I think perhaps one should say it has to do with unexpected simplicity, where one imagines that things are going to be complicated but suddenly they turn out to be much simpler than expected.

He revealed his tile set.

I feel that the pattern of Fig. 4 has a certain aesthetic appeal.... One often becomes fascinated by a problem even although it has no particular relevance to things one may be doing. And when one is fascinated, the internal aesthetics of the thing will drive it along. Sometimes it turns out that what one has been doing can be used for something else, but sometimes not at all. That is one of the nicest things about mathematics.... So often it must be the case that people just do things for their own sake, merely for the enjoyment of doing them, and this enables progress to be made that could not be made otherwise.

Roger gave himself over to the "pleasure one feels because one is searching for truth." That singular joy overrode his personal relationships. He remained infatuated with Judith, but that desire was subsumed into his pursuit of the most promisingly beautiful mathematical solutions.

14

FAITH

"It is always possible to continue forever."[1] So wrote famed *Scientific American* mathematics columnist Martin Gardner about Penrose tiles. But the relationship that produced them was finite.

Roger and Judith's world lines were diverging.

By late 1974, Judith was beginning to regain her happiness.

> I'm feeling much more cheerful recently than I have done for some time. I know I'm moody anyway but that's a fairly superficial thing affected by the weather, the time of the month (or used to be when I had a time of the month), how tired I am, if my feet hurt, if I feel fat, etc.! But when I say I feel happier I mean in a deeper sense.—I don't know why. Since I went into hospital . . . I feel rather as though I've been walking through a tunnel, a maze of tunnels. I'm still underground but I think I can see the light.[2]

She broke up with Colin and kept Roger at more of a distance. She began a new relationship with a man named Richard Pike and, in 1975, moved in with him.

Roger was confused and angry. He had assumed her unwillingness to marry him reflected her general disinclination to commit to anyone. This assumption fed his belief they were united in a higher purpose that transcended normal human love and attraction. He could no longer tell himself this story.

He withdrew into jealousy and disappointment. She accused him of emotional blackmail. He insisted he was helpless to behave any other way.

"If I seem to be threatening to withdraw myself from you if you don't express sufficient affection, it is not a blackmail thing at all. It is that I have to come to terms with my own feelings within myself. If I feel that, for one reason or another, a deep and meaningful relation with you has become impossible, then a radical (and painful) readjustment of my own attitudes in relation to you is necessary," he wrote. She had chosen Richard's world over his, he reminded her. She was the destructive one, not him. "I was attaching a greater aesthetic value to the deep eternal truths, the unchanging and unchangeable structures and concepts, be they the works of man or the works of God; whereas you had seen more to be valued in the fleeting intangible impression of the moment, the smell of a flower that catches one unawares, the very fragile ephemeral beauty of the flower itself, that is soon gone never to reappear in that unique form again."[3]

Judith couldn't quite believe he would give up so entirely on their friendship.

If it is true that the magic we shared, the inspiration I was for you and the mentor-mathemagician you were for me, if that must really disappear then it's a tragic loss. And not just a loss to my personal happiness or yours but a much more significant loss—to philosophy and mathematics perhaps, to our own faith certainly, and somehow a loss in the "rightness" of things that I'd always at least half believed in.

Perhaps it is the beginning of a new phase in your work, perhaps there'll be someone else or you'll find a new way to inspire yourself. Maybe my role was essentially temporary and is not right for whatever you'll be doing next. I don't know. That's the awful thing about all of this—I am out of my depth, I don't know the answers. I just wanted to convey to you, without in any way blaming you, how completely lost and powerless I am in this situation. That I feel sad at the loss and frightened by the thought that I could have caused it—frightened that I might have had that power but not in my control.

Judith's power over Roger wasn't hers to define. He had decided she was his muse without regard for how she viewed herself. He had crafted the narrative in which they conquered the universe together. Her own creativity and agency were not a part of the equation. When his expectations pushed them to the breaking point, her lack of choice in the matter was laid bare.

Roger wrote Judith a farewell in the form of a fable he titled "The Wolf and the Squirrel." He hand-delivered it to her in an envelope on which he wrote, "One extra fable for Judith... (I hope it's not too silly!)—Roger."

The Wolf and the Squirrel

One day a wolf was walking through the woods in search of prey, when he came upon a young squirrel whose foot had got caught in a crack in a hollow tree. The wolf took pity on the squirrel, for he was much moved by her beauty and helplessness. So breaking off bits of the tree with his teeth he was able to free her unharmed. The squirrel was grateful to the wolf, but was a little frightened that he might decide to eat her. So she climbed the nearest tree and watched the wolf from a safe distance.

She realized how fine the wolf looked with his sleek fur and well-proportioned body, and how well-suited he was for his job of hunting. For many weeks afterwards she would sit and watch him admiringly as he hunted sheep or rabbits or rats. The wolf was aware of her gaze and strangely gained in strength and cunning from the knowledge that she was there. But then sometimes he would look enviously at the squirrel when she leaped from tree to tree and was joined by other squirrels.

He began to think to himself: "How I wish that I were a squirrel! Then I could join my friend among the branches and live off nuts as she does; so I would not have to kill other creatures in order to live." But the wolf knew that this was a dream. For his stomach was not of the kind that could digest nuts; and he knew that he would starve without meat. Presently he noticed that the squirrel had moved to a more distant tree to watch him, and was watching him less, spending more and more time with the other squirrels.

The wolf was moved to a certain sadness. He knew that he had to resume his hunting. But though he continued to catch many animals, it was somehow different—his thoughts were still among the trees.

Drawing by Roger Penrose accompanying "The Wolf and the Squirrel."

Judith disappeared into a happier world where he couldn't follow. The ever-present burden of his talents and ideas pulled him in another direction, piling up around him, walling him off from the trifling pleasures of ordinary romance and intimacy.

A friend and colleague, fellow British mathematician John Conway, saw more potential in Penrose tiles than Roger himself did. John was famous for inventing the "Game of Life," a mathematical game of "cellular automata" with as much popular appeal as Penrose tiles. He devoted his considerable knack for capturing the mysticism and poetry of mathematics to bringing Penrose tiles into the public imagination. He coined the terms *kite* and *dart* for the prototiles and came up with nicknames for common tile groupings: the king, queen, jack, and ace; the cartwheel; the sun and the star; the wormhole and the spoke.

John cut out his own Penrose tiles and spent more than a year playing with and analysing them before connecting with Martin Gardner at *Scientific American*. Gardner's *Mathematical Games* column was the touchstone for recreational mathematicians around the world. Professional mathematicians read it, as did non-specialists who just loved games, puzzles, optical illusions, patterns, tricks, and magic.[4]

With John's assistance, Penrose tiles made the cover of *Scientific American* and quickly became an essential component of mathematical folklore. Even people who found most math alienating had their breath taken away by these simple, satisfying tiles and the infinitely complex patterns they created. For elite mathematicians, "unexpected simplicity" generally came wrapped in a cloak of esoterica. Penrose tiles laid it bare for anyone to see. People were in awe of the tiles and their creator.

Roger's fame grew without his really trying. His head was elsewhere, still negotiating his new relationship with Judith. He wrote,

> I wanted to say something more about the "Wolf & Squirrel" fable....Perhaps your interpretation of the wolf's carnivorousness contained a strong element of truth that I had not (consciously) intended. And our differences in interpretations may contain a germ of something of significance to the kind of relationship that we have had. To you, it seems, the wolf's carnivorousness represented male sexuality in some form—and you didn't see quite that it was "me" or that I could feel that way. But to me, the carnivorousness represented something related to my own creativity. Perhaps that burning inner drive of ambition that does not allow me to be satisfied by elegance alone; or perhaps it is something in me that goes along with that

creativity, something that makes quite ordinary tasks almost impossible for me to do—for I need that feeling of stimulation for whatever-it-is to seem worthwhile at all.[5]

He had written to her before about how his sex drive had become sublimated into his creative drive, a conflation that came out in his fable whether he acknowledged it or not. It also manifested in his subsequent hunt for another woman to appreciate and inspire his work.

He was not the sleek predator of his fable. He tried to recreate the dynamic he'd had—or imagined he'd had—with Judith with a graduate student at the maths institute, but she rebuffed the arrangement he proposed to her.

You had said you would like to be able to talk about twistors to me—and also that you'd like a chance to see me. Well, I was very flattered by this—probably more on the first count than the second, because I was trying to "find my feet" in the academic world....I agreed on the basis that it would be fun, and possibly fruitful, for you to talk to me about your work. And I certainly have found it interesting and very stimulating—probably in particular because I've seen first-hand what "research is all about"—at least for one very successful person. I am delighted you've found it helpful....Now it seems that you really require more of me—a keenness not confined to talking to you on this kind of level, but also to seeing you as well. Well, I do enjoy seeing you, but also, I'm afraid that I'm not the person to give you what you appear to be looking for.[6]

Roger didn't understand what he was doing wrong. The conflation of work and intimacy felt natural to him. He habitually turned to Judith for advice. Judith had ongoing medical issues and was working on a master's degree. She had less time and energy for Roger.

I want to apologize for being so off-hand on the phone on Friday. I really wanted to write you a decent reply to your lovely, moving letter but apart from time shortage, I don't really know what I want to say. This week has been quite ghastly—I had the x-ray yesterday—it was even more painful than I expected and lasted far longer—they did it on TV rather than just "stills" so it went on being agony for some time. I'm afraid I wasn't at all brave—I screamed and kicked and begged them to stop and it didn't help to know that if only I could relax it wouldn't hurt so much.

I'm still not right—I didn't give my lecture today. I've stayed up 3 nights preparing it, but I hadn't got it typed anyway—I went in and arranged to give it next week.

Roger—I am sorry about everything—I don't know what's wrong with me these days—I seem to be deliberately isolating myself. I don't know if it's the exams or work in general, but I feel I can't cope.[7]

Roger, disappointed Judith didn't have the energy to counsel him, saw no reason to visit with her at all. "The trouble has been that with my own life as regards women/work/inspiration in an unsatisfactory state, I would find difficulty in seeing and/or communicating with you unless I could view it as in some way helping to improve that state. A narrow view, perhaps, but with pressure of work as it is, a mere pleasure in seeing or communicating with you would not be sufficient without some such excuse!" he told her.[8]

New ideas were piling up in his mind and tumbling out in lectures and writing. He published a novel theory about using nonlinear gravitons to incorporate general relativity into twistor theory. He delivered a landmark lecture to the International Astronomical Union about his (unproven) cosmic censorship theory, which argues that *naked singularities*—singularities not hidden behind event horizons—cannot exist. He wrote a chapter titled "Is Nature Complex?" for *The Encyclopedia of Ignorance* anthology—he still yearned to prove that complex numbers underpin reality. He appeared to move effortlessly from theory to theory, leaving his colleagues shaking their heads with admiration.

"The sheer brilliance of Roger! He's one of the handful of people I've met in my life I would apply the word 'genius' to. If you forget twistor theory, forget his ideas about quantum mechanics, about everything else and just look at his contributions to classical relativity—I would say he's the most important person who worked on the subject after Einstein," said Lee Smolin, one of the world's most influential theoretical physicists.[9] "His focusing on causal structure, conformal completion, spinorial formalism, and using spinors to get at the geometry of null rays, which grows into twistor theory. And then all the uses Roger put that to, into so-called peeling theorems that allow you to disentangle gravitation radiation from other variables that don't radiate. All those plus the singularity theorems, even though one of those papers was joint with Stephen. I mean Stephen was a very good physicist, but Roger was the creative source."[10]

Without a muse, Roger mistrusted his ability to remain creative. He yearned for another woman, another Judith, to inspire and appreciate him.

In late spring 1976, he made what he felt was his first significant advance in twistor theory since he and Judith stopped meeting regularly—a method for "growing new twistors."

Over dinner with a group of young scholars at the university, he excitedly recounted his new ideas to a classics student who happened to be seated next to him. She responded politely but without great interest and returned to the general conversation. He burned with such chagrin, even he recognized his own reaction as disproportionate. "I suppose ... women are still important to me in my work—even though I am much more independent of that now than before. And I am still sensitive to women's reactions to my work—even if it's quite superficial," he wrote to Judith.[11]

With every failed attempt to impress someone new, he felt the loss of Judith more profoundly.

I said I'd comment on the question of feelings about religion. It's strange but my relation with you had almost changed my views on such matters. Whereas I could always (like Einstein!) believe in

Spinoza's impersonal "God" (if I have him right), I had, with you, almost come to believe in something more.... But then your relation with Richard dashed that feeling in me—tenuous as it was. I suppose, more than anything, it was the almost "religious" sense of rightness that you said you felt you had found in your relation with Richard that I could not take. That you could be moved to love by whatever mechanism it is in people that makes them feel that way, I can accept. But that it should be God's will—that makes no sense to me at all. For if I am to see anything of meaning in the idea of a God who concerns himself with personal feelings, it would be by searching within that special relation that we have had together. And I cannot see why—from my own personal selfish viewpoint—that same God should, seemingly wantonly, wish to destroy that relation.[12]

Any God who wasn't on board with him and Judith being together couldn't be much of a God at all. Without her, his successes felt diminished and unsatisfying.

His work with Wolfgang Rindler on *Spinors and Space-Time* had nearly reached a standstill. Wolfgang travelled to Oxford so they could breathe life into their aging manuscript.

I had hoped that we could just rattle through it, but no. We ended up doing 1/4 to 1/3 of what we had intended for the period.... I feel a great pressure of work on me most of the time. But the problem is not that I do not find time to "rest," so much. It is more a problem of lack of incentives for work that I do not want to do. When well-motivated I can work efficiently, need less sleep, am cheerful and can find time to enjoy myself. When unmotivated I can stare blankly for hours at a mess of papers on the floor. I feel tired and sleep longer, am depressed and can find no time to enjoy things.

It was you...who had kindled a fire in me that had breathed life back into the book-writing project. That I could interest you with what I wrote—that meant everything to me. It made the whole project exciting to me again. I feel a fool and vulnerable. I wonder what dedicating my work and feelings to you really meant. Yet I know that it did (even does?) mean a lot to you—even something irreplaceable. I know that I could not have been inspired by you if I had been that mistaken about your feelings. Yet the turning away from me is real—and it was you who took such pains to emphasize its reality. Though all the time you tried to tell me that Richard was taking from me only things that I did not have in the first place, I'm afraid that I am not able to really accept that. To accept that would be much worse, for it would mean that I really was mistaken about your feelings; that my source of inspiration had been based on a fallacy. But my inspiration was based not on what you said, but on what I saw in your eyes. I would rather leave it that way.[13]

In these moments of reflection, Roger was honest with Judith and with himself in ways he had not been before and would not be after.

Years earlier, Roger had stormed out of a meeting with Joan and a therapist, indignant over accusations that he'd had an affair. Now he finally acknowledged it.

We, having that strange yet in its way wonderful affair—not that it would be accepted as an "affair" in the usual sense of the word, of course (and I suppose that for you that was the "point"), it was not, as far as I recall, even accepted "officially" by you as being an affair. Yet, in emotional content that is what it was—and for that aspect of things I have no regrets whatever. Apart from the fact that it was enjoyable for me, it held me together during a very difficult stage in my life and even turned me right around from a time when research was for me impossible, to a time when ideas flowed as they had never done before.[14]

This moment of clarity led not to greater self-recognition and understanding but to a new wave of frustration with Judith.

"I feel a bitterness, in a way, that it never was allowed to become a 'technical' affair. I sometimes wonder now what it was all about. There is something in me that feels 'conned' by it all, I suppose. How could you have drawn those things from me if it had not meant something deep to you."

Even in the midst of such harsh self-realizations and reprimands, he could not keep himself from trying to talk to Judith about his work. One of his most promising and productive students, George Sparling, had taken a postdoctoral position in Roger's group and had become a major champion of Roger's most cherished ideas.

He's become something of a fanatical twistor theorist, producing endless successions of original ideas and detailed calculations, and absorbing and re-explaining relevant mathematics, being very helpful to students, and generally spreading his enthusiasm to all around. I'm sure my research group has benefited enormously in many ways from his presence. . . .

It is perhaps fitting that on Friday evening, when a group of us were having dinner eating pizzas (I was filling in time waiting for the bus that the kids were coming to Oxford on), George kept mentioning something that was disturbing him about how things were fitting together—or failing to fit together about the twistor programme. At first, I didn't see what was worrying him. But then I realized it was a very significant point. But it suggested something different to me which I feel, now, may represent a significant advance in the whole theory. I'm quite excited about it all. . . . There are new and wonderful things that I could never have guessed. Sometimes I find it a little frightening. I don't know if I can really explain. One wants one's ideas to work out, of course, but there is something awesome about it all if one is really seeing some new corner of "God's design." I would really love to be able to explain my thoughts to you—including the details of the mathematics. I need to be able to share not just my hopes and excitements but my frustrations and my fear too.

He couldn't let go of twistor theory, and he couldn't let go of Judith. In a single letter, he both lamented the impossibility of recreating the magic they once shared and attempted to do exactly that. He wouldn't take no for an answer—from her or from the universe.

She trod carefully in her responses.

Why can't people accept that feelings change? We are like children wanting the security of permanent unconditional parental love and if the love changes in form we feel wronged or that we must have done something wrong to be punished like this. I want to try and help you understand why I don't want a sexual relationship with you when I don't really understand myself. I just feel it wouldn't work. I'm writing in the present tense but really, I'm talking about the past. And here's where our "distortions" of memories differ. I think you would claim that I did want you in that way but, for various complicated reasons, repressed it. It's true that I did sometimes physically respond to you and "want you" in that very immediate sense but I don't think I ever really wanted, emotionally wanted, to have an "affair"—I mean to go to bed with you. I wanted to want to, I wished it could be my great romantic dream come true, but I think I always knew it couldn't be—not because of morals or inhibitions but just because my emotions about you weren't right.[15]

Roger and Judith's letters became less frequent and more predictable. Birthday cards and hurried relationship news. Occasional references to new ideas and theories. Judith moved, remarried, got divorced again. Roger pushed on with twistors, convinced his most important discoveries still lay ahead but worried they might never happen without his muse.

15

THE GAUGUIN DECISION

Born in Liverpool in 1948, Salley Vickers trained as a Jungian psychotherapist, specializing in helping people get past creative blocks in their work. She married a British physicist named Martin Brown, who studied quantum gravity. Martin had studied under Dennis Sciama and, in the late 1970s, joined the Center for Advanced Study in Austin, Texas, where Roger had worked in 1963. Salley left her teaching position at the Open University to follow him.

"When I came out to Texas, I was not very happy. I didn't have a job. I didn't terribly like the society. Texas rather threw Martin and I together in a way we formerly hadn't experienced. We were very, very different people. We married very young. We had children very young. He didn't really want children, so that was an issue. The differences between us came to a head," Salley said.

Her marriage was breaking down, and Salley found herself utterly alone. "All of my cultural life, my working life, and all my friends disappeared. There weren't people I could be friendly with. I didn't have anyone to confide in. I just had the children," she said.[1]

She knew of one person who might understand what she was going through. She had met Roger only briefly on two previous occasions. She knew about his separation from Joan and sensed he would empathize with her situation.

> This will seem an odd communication, and indeed I feel uncomfortable making it, because you will probably have to pause to recall who I am....I have felt an inclination not so much to ask your advice (because I am sure you are the kind of person who would be reluctant to proffer advice) but more to talk to you about certain things that were disturbing me because I found you especially empathetic....Martin and I are parting. A lot of things have been exacerbated by our stay in Austin, and removed from the supports and habits of our habitual life I found certain things almost unbearable....I feel a simple desire to talk about the children and what is best for them with someone who has experienced the unhappiness of breaking up a family.[2]

In the late 1970s and early 1980s, Roger regularly visited colleagues in Dallas and Austin and at Rice University in Houston. He and Salley met up quietly—gossip spread quickly among physicists, and neither of them craved the attention. When their friendship became an affair, they became even more careful and secretive.

Neither a scientist nor a mathematician, Salley still understood the creative value of Roger's work and the forces that motivated him. That was plenty for him to slot her into a familiar relationship. "I could connect with Judith through the actual mathematics. Salley didn't understand the material itself, but she understood the drive behind it better than anybody else," he recalled.[3]

He constructed a role for her that was even better and more powerful than the one he had created for Judith.

The workings of intuitive thought are mysterious. It is strange how often I have felt that I had "known" the answer to a problem, sometimes even years before the key mathematical ideas had become available to me. So I didn't really "know." And sometimes I have found myself to be mistaken about some essential point—which had held me up—yet, at a deeper intuitive level, I was right all along! Mysterious and odd—yet not mystical. You can help me to understand. You have an instinctive magical understanding of such things. That is very valuable. I have come to be convinced that in you there is the opportunity that I shall never encounter again. I believe that you are virtually unique in possessing both the intelligence and emotional understanding to bear with me.[4]

Salley appreciated his enthusiasm. She was perceptive, well read, and self-aware enough to know his ardour had more to do with his own aspirations than her actual qualities. For a time, she willingly played the part.

I began to see that I was really rather a muse. Although he appeared to be absolutely nuts about me and very, very passionate, I didn't really believe it. I didn't believe it was really about me. I just remember he was riveting and enchanting when he talked to me about these ideas. I felt I could see them. I could intuitively feel the truth of them. I'm not in any doctrinaire way religious, but I do have an interest in the kind of religious sensibility. And Roger is one of the very few mathematicians I encountered in that world who thought easily in those kinds of terms. Not about God or anything like that, but in the sense of something beyond the limitations of this world. I wasn't flattering him. I wasn't doing it to sort of please the man who was my lover. I was doing it because it was truly enthralling.[5]

Roger accepted an associate professorship at the University of California, Berkeley (which didn't require him to give up his Oxford affiliation). He and Salley found a house together there, immersed in the beauty of Northern California and enjoying and enduring the chaos of their five boys coming and going as joint custody allowed.

A pair of incorrigible insomniacs, they'd put the kids to bed and then stay up late into the night, Roger expounding on twistors and aperiodic tiles. She had a self-awareness and capacity for analysis that helped Roger open up not only about his work but about his sense of his own specialness and purpose.

> There was a sort of delicacy about Roger and a tentativeness. One of the things that came up quite a bit when I was with him was the work he did with Stephen Hawking. He was too modest to actually say it outright, but I think he felt Stephen had got the lion's share of the praise for the work they did, and that Roger had not been adequately acknowledged. He minded that, but he didn't come out and say it. There's a sort of an odd mixture in him of an absolute conviction that he's right about things, at the same time an over-modesty.[6]

In daylight hours, he improvised piano pieces in the style of Mozart and made custom versions of Penrose tiles for Salley's sons. She noticed, though, that he was distant with his own children. "Although Roger was very good to my children, I felt his relationship with his own children was patchy. When they were visiting, he tended to leave the interactions with them to me. I was especially concerned by what I discerned as signs of mental ill health in one of his sons," she said.

Roger showed no capacity or interest in dealing with such issues. Salley observed him without blame or judgment but increasingly questioned whether she and Roger really saw children the same way. "This is a familiar pattern with great artists and great geniuses. I mean, it is what they do. They're ruthless—in the nicest possible way in Roger's case. They have to be. They've got something they feel is unique and only they can uncover. I understand it. But I suppose, because of the difficulty I had with my husband over my own children, it did set up alarm bells. I wasn't confident that if I made the move to live with him, my own children wouldn't suffer."[7]

"It was me who suggested he had a 'Gauguin complex.' That was absolutely my term for it," she said. The nineteenth-century French neo-impressionist Paul Gauguin abandoned his wife and five children to pursue painting.[8] His artistic impulses overwhelmed every other responsibility.

With the stint in California over, Roger returned to England. A few weeks later, following a trip to Ireland with his sons, he wrote Salley a jumbled letter justifying, confessing, and attempting to demonstrate his paternal bona fides while acknowledging other priorities.

I didn't get back to Stanmore until nearly midnight on Sunday—too late to return to Oxford. Staying until Monday evening at Stanmore enabled Joan to stay another night with her boyfriend, me to take Christopher to his 6th-form college on his 1st day (I think that he was quite nervous about it and was very glad to have me take him), and gave me a chance to fix Eric's little telescope to his big one as a "finder." I had to make a special wooden attachment—something I had been meaning to do for some time, but had not had the chance before. E. had been really marvellous on this trip, doing all the map-reading, often while the other two slept. He really deserved something extra.

He resisted the idea he had a "Gauguin problem," while confirming it more and more with each sentence he wrote. "I have never met anyone else with such a magical ability to read me as you have. Yet I fear that there are factors and complications that you have not always appreciated. I refer to...the awesome responsibility that the possession of special talents brings with it."

He had overcome his self-doubt over the value of his work. In its place was a conviction he must pursue his ideas regardless of the cost. His drive to conquer the universe had crystallized into a matter of cosmic necessity.

I refer also to a certain destructive impulse that I have noticed in myself when...those responsibilities come into conflict with other responsibilities (e.g. to family or students). This destructiveness seems to result from an anger towards one's parents—or perhaps to a God—it says, "Damn you! You gave me these talents and you loaded me with these responsibilities, yet you failed to give me the abilities or the circumstances to cope with them. So to spite you I shall let them wither and die!" Yet in my feelings towards those talents is something akin to the love that I feel for my children. Indeed, it is something so painfully close to that, that I cannot prevent the tears from coming to my eyes as I write this....I do still feel that I did not have a real choice when I left Stanmore—and that without the help of another wife (or effective wife) I could not have properly looked after my three children in Oxford while holding down my job....I can, even now, see no practical alternative to the "Gauguin" action that I did take.

He understood the price. Salley had communicated it to him, and he acknowledged it.

Where I feel that you have hit me with a deep truth (and please don't take this as a complaint. I am truly grateful to you for it) is in making me realize that the children feel deserted by me. I can see it particularly with Toby....Perhaps it was from Toby that you obtained the impression of me that I had made a "Gauguin decision." Your instinctive affinity for children must have made it immediately obvious that he felt this way—while my blindness and need for self-justification had served partially to hide the nature of his feelings from me.[9]

For Toby, those feelings stemmed not from Roger and Joan's separation but from his father's obsession with mathematics and physics. "It seems like his biggest love affair was with his work," Toby said, later adding, "I think he had great insight into me in many ways, but at the same time, there was a great distance in how that was communicated."

Roger, who frequently lamented his own father's intellectual fierceness and emotional distance, could not find a way to do things even slightly differently with his own sons. He sensed he should feel guilty but saw his children's estrangement as a necessary cost. Salley respected his view but didn't share it.

"I understand his desire to duck the quotidian demands of family life. I have never blamed him for this. But for me, children have always been paramount. I parted from him with regret. Although I did love him...there were things that were going to be more important to me. All ways of life have a cost. And I chose that particular way," she said.[10]

Alone again, regretful and relieved, Roger threw himself into his work.

16

UNDERSTANDING

One Friday afternoon in 1987, David Deutsch entered Roger's office at the Oxford Mathematical Institute. He surveyed the space. A huge blackboard entirely covered one wall. A jumble of chairs and desks filled the spacious room. David plunked down in what looked like the most comfortable chair and waited.

David had recently finished a physics PhD at Oxford's Wolfson College. Two years earlier, Roger had helped him publish a paper in the *Proceedings of the Royal Society of London*, in which he proposed the idea of a universal quantum computer. David's thesis advisor, the omnipresent Dennis Sciama, had prevailed upon Roger to take David on as a postdoctoral fellow.

"I wanted to come back to Oxford, but the only way we could arrange this was that Dennis persuaded Roger to take me under his wing, even though I wasn't at all a mathematician and wasn't at all competent to participate in this group," David said. "It was all either twistor theory or relativity at a level way above anything I knew about."

Other postdocs and graduate students filed in, eying David with sympathy and curiosity.

"You don't sit there. That's Roger's chair," someone finally said.

"I jumped up and moved to one of the uncomfortable chairs," David recalled.[1]

Roger's Friday afternoon meetings had become the stuff of legend on campus: weekly informal, creative discussions where extremely wild ideas were welcomed—as long as you could back them up with reason and science. The gatherings didn't have the same edge as Lionel's parties at 47 Lexden, but there was just as little doubt about who owned the room.

Officially, these were meetings of the Relativity Group. Many of Roger's students called themselves "Twistorians." Under his supervision, they produced a typed and handwritten zine called the *Twistor Newsletter*. Most issues included Roger's own essays and diagrams, photocopied and stapled together alongside those of grad students and postdocs.

David never ascended to Twistorian status. His first Relativity Group meeting was also his last. Instead, he and Roger began one-on-one conversations about quantum computing and human consciousness. "There were one kind of argument that said, 'We don't have consciousness,' and another kind that said, 'We do have consciousness, but it's supernatural.' And both of those things Roger and I violently disagreed with," David recalled.

Roger's curiosity about consciousness came from many places— the extreme mental feats of his father and brothers, the speed of decision making in racquet sports, the human ability to transcend Kurt Gödel's incompleteness theorem, and his own capacity to discover new mathematical insights.

Many physicists believed consciousness played an important role in quantum mechanics. According to some interpretations, a quantum object (or system) can exist in many places or states at once. An electron simultaneously spins clockwise and counterclockwise. A photon passes through two openings at the same time. In quantum computing, a statement can be both true and false. When a conscious observer "measures" a quantum system, it "collapses" into a classical state of one thing or another.

Roger's scepticism of certain interpretations of quantum theory reflected his more general sense that the entire field was catastrophically incomplete. He considered it a reasonable placeholder theory until something without so many gaps came along. He held special contempt, though, for the idea that a conscious observer was required to trigger the collapse of a quantum system.

He suspected physicists were getting things backwards: What if consciousness didn't cause quantum collapse, but instead quantum collapse caused consciousness? He discussed his half-formed ideas with David and periodically shared them with favoured members of his Friday afternoon group.

Oxford student Tristan Needham had read all of Roger's papers from the 1960s on gravitational collapse and trapped surfaces. His own visual sensibilities and intuitive grasp of physical geometry naturally drew him to Roger. "I basically fell in love with Roger's mind when I was twenty years old, my second year as a physics undergraduate," he said. "I already had this incredible respect, reverence for Roger before I ever met him in person. I was just blown away immediately."[2]

In 1987, he applied to join Roger's group. Roger had funding for only two students per year, and competition was fierce. "My private interview with Roger was terrifying," Tristan said. "He was all business, grilling me on technical aspects of theoretical physics." Tristan survived and became one of the group's most dedicated Twistorians.

Friday afternoon meetings typically drew about twenty people, who sat on tables, chairs, and the floor. Most brought brown-bag lunches as fuel for the multi-hour sessions. Roger's informality was refreshing and unusual. "It was in stark contrast to when we used to go to Cambridge to visit [Stephen] Hawking's group. There they would be going down the halls practically clicking their heels and saluting," Tristan recalled.

Roger stood at the floor-to-ceiling blackboard, drawing light cones, Reimann spheres, Penrose diagrams, and the occasional impossible object. His hands and face became streaked with multicoloured chalk dust as ideas took form on the board. "Roger would present something almost every time. Those meetings were a chance to really see him just thinking out loud and in particular to see these beautiful drawings that he would do in real time with coloured chalk, which I have emulated to this present day," Tristan said.

Twistors, spinors, general relativity, Hawking radiation: Roger was ready to draw anything people wanted to talk about. Tristan recalled,

> I was just so impressed to see how almost any problem could be resolved using geometry. I loved geometry already, but to actually see the power of it in the hands of a genius like him to explain even things that seemed to have no connection to geometry whatsoever. He was able to bring geometry to bear and to draw these beautiful pictures that would explain everything. In high school and college, I felt like I was often the brightest person in the room. I had a fairly high opinion of myself. But when I got into Roger's group, I suddenly realized I was the village idiot. Here were these unbelievably brilliant people from all around the world, Rhodes scholars and so on. And oh my God, that was quite a wake-up call. That feeling never went away.[3]

Students reacted to Roger the way Roger reacted to seeing Saturn floating in the night sky: their transformation was emotional as well as intellectual. "I feel that I could see aesthetically Roger's greatness in a way that I don't think almost anybody else could," Tristan said.[4] "I think maybe he saw in me that I saw in him what he cared about himself."

Roger had tenure, abundant honours, and a new generation of researchers building on his most cherished ideas. Both volumes of *Spinors and Space-Time* were finally published. The acknowledgements included the line "Especially warm thanks go to Judith Daniels for her encouragement and detailed criticisms of the manuscript when the writing was going through a difficult period." This was the only time Roger ever publicly acknowledged Judith's contributions to his work.

He had an enviable career, but he still felt restless and incomplete. The universe remained unconquered. Twistor theory still hadn't reconciled gravity and quantum mechanics. Meanwhile, theoretical physicists were rallying around string theory, leaving Roger once again at odds with the scientific mainstream. "String theory really took over," said Paul Tod, a long-time friend and colleague of Roger's at Oxford. "Everybody who got a faculty appointment in theoretical physics since the eighties has been in string theory. It really was hegemonic. It's clever stuff. It's hard to do. And it had absolutely no experimental support. Levels of bullshit crept into the grant applications, but people ended up believing it.... Theoretical physics becomes very arrogant. And I think that offended Roger."

Roger was equally offended to find Stephen Hawking had once again "fallen in with the wrong crowd," this time on the question of whether black holes destroyed information. It was obvious to Roger they did: matter and energy disappeared into a black hole, which then boiled away into entropic Hawking radiation, leaving no record of cosmic history.

Most physicists considered the indestructibility of information essential to the very idea of cause and effect. If information could truly be destroyed—rather than merely transformed, scrambled, or made inaccessible—there was no way to reconstruct the past or predict the future. The position and state of every particle in the universe results from the conditions that existed a moment earlier. What happens next depends on what things are like now.

The destruction of information in a black hole would create effects with no causes. Physicists called it the *black hole information paradox*, but Roger saw no contradiction. "I can see why they say that, and I can see why I don't say that. One of the most fundamental principles of quantum mechanics is this thing called 'unitarity.' And with unitarity, you can't lose information."

Unitarity relates to the Schrödinger equation, which describes how the probabilities of quantum wave function evolve over time. All these probabilities must "add up to one," meaning all potential futures are baked into the present moment.

Without unitarity, quantum physics was in trouble—which suited Roger just fine. "If these horrible singularities swallow information, if you take all the Hawking evaporation, and slog your way through everything you can think of, and you still lose information, you break unitarity. People regard this as a huge paradox. I don't, because following the Schrödinger equation is not what the universe does."

The universe wasn't like that. Information could be destroyed. Consciousness arose from quantum collapse. Quantum theory was incomplete. Sting theory was bunk. The universe didn't start with a singularity. Roger had a growing collection of deep disagreements with mainstream scientific consensus.

He was also edging toward challenging the closest thing physicists had to a gospel truth: the second law of thermodynamics.

The second law says any closed system only ever becomes more random and unstructured; entropy only increases. Information-destroying black holes might skirt around the inevitability of the second law, not exactly reversing entropy but possibly allowing a kind of reset button for the universe.

Before he could focus in earnest on that possibility, he became distracted by how wrong everyone seemed to be about consciousness and human understanding.

BBC radio aired a conversation between Marvin Minsky and Edward Fredkin, two pioneers of artificial intelligence. "Marvin and Edward were saying two computers could communicate more ideas to each other than the entire human race ever communicated in the time it took the scientists to walk across the room," he recalled. "I thought, 'I know where you're coming from, but I just don't believe you.' Computers don't *have* ideas, let alone communicate them."[5]

Minsky and Fredkin's conversation crystallized a lifetime of back-of-the-mind questions about Rock Paper Scissors, logic puzzles, aperiodicity, and his own flashes of inspiration about singularities and twistors. The human brain did too much too quickly to run on algorithms. Gödel's incompleteness theorem had become a touchstone for him—proof that human understanding was non-computable.

He felt he could write a book about everything Minsky and Fredkin got wrong.

As Salley had foreseen, Roger found it increasingly difficult to maintain a close relationship with Christopher, Toby, and Eric. They often came to Oxford on the weekends. Roger took them to classical concerts, found a tennis court where they could play, and invited them to draw fantastic creatures on the chalkboards in his office. Still, he had trouble sustaining his ability to engage with them.

On a visit to David Deutsch's house, Roger noticed a Betamax VCR hooked up in the small living room. He offered to cook supper for them all if David would let his sons come over and watch a movie. Roger invited Tristan as well. "I think Roger honestly had a hard time dealing with his children. Roger recognised that I was sort of socially adept and able to get along with different people. He would bring me into the mix and invite me to dinner, and I would be like a buffer," Tristan said.[6]

Other members of the Relativity Group caught wind of these movie nights, and David's small home soon became crammed with Twistorians and other physicists every Saturday evening. A graduate student rented a horror film or Hollywood blockbuster. Tristan drove Roger, his sons, and bags of groceries to David's house.

David opened the door one Sunday to find Roger and Tristan holding enough ice cream to feed the crowd. "My tiny one-person fridge had an even tinier freezer at the top. The boxes of ice cream on the face of it looked about twice as large as the freezer," he recalled.

Some of the world's most promising scientific minds crammed into the kitchen, trying to solve the problem.

"Maybe we keep some of it in the fridge?" one suggested.

"Maybe we should eat the ice cream first," said another.

"Well, let me have a look at it," Roger said.

The crowd looked on as he shifted items around the freezer and fridge. It dawned on David exactly what they were witnessing. "Over about fifteen seconds, we went from one state of mind to another. First, we were sure that he couldn't possibly fit it all in there because we couldn't. Then it looked as though he was making headway. And then we had a funny moment when we looked at each other and realized we had just asked the world's leading expert on packing things to pack the ice cream into my freezer. And sure enough, he managed it. It was an impossible task, but he managed it."[7]

As blood and gore spattered David's television screen, Roger kept a journal handy. They frequently paused the movie to complain about plot holes and to float ideas about quantizing gravity or artificial intelligence. "We would stay up to crazy hours, talking until three in the morning with these crazy movies and conversations about quantum mechanics and relativity and so on. Roger would have his notebooks open, and I would see him doing these amazing drawings. It was a wild time," Tristan recalled.[8]

Nearly seventy years had passed since Margaret studied medicine at Cambridge, and women were still a rarity in elite academic circles. Roger had one female doctor of philosophy student, a talented twenty-one-year-old mathematician named Vanessa Thomas.

"I adored undergraduate courses in general relativity and cosmology. They completely brought me alive," she recalled. "When I asked for recommendations on where I could go for graduate study my tutor at the time said Oxford and the only person worth studying with is Roger Penrose. So off I went, and made an application."[9]

He was famous, respected, and charming. She was brilliant, young, and appreciative. She didn't join the Saturday night horror shows at David's house, but she and Roger attended lectures and other academic events together.

"There was nothing secretive about that," Roger said.

They did have a secret, though.

"Toward the end of my DPhil, I would go round to Roger's home in the evening, and Vanessa would mysteriously be there. Nothing was explained or discussed at all because she was still officially a student and so it was all wildly inappropriate. But, you know, I put two and two together," Tristan said.[10]

She felt confused to be involved with such a prominent figure. "At twenty-one, I was naive. How could you not be at that age?" she said.

Roger had a friend who lived in the village of Over Wallop named James Dalgety. Neither could remember where they met or how their friendship grew. They shared an obsessive fascination with puzzles of every type. James had collected thousands and thousands of puzzles—rope-and-ring challenges, jigsaws, burr puzzles, wooden puzzle boxes. His collection spanned centuries and included sets of Penrose tiles, as well as copies of Roger's tetrahedral 3-D jigsaw. Many of the handmade items in his collection were beautiful works of art as well as clever brainteasers.

His cottage became an invitation-only puzzle museum, open to only a handful of puzzle enthusiasts who could not only appreciate the items in his collection but also solve them. Roger and Vanessa frequently snuck away from Oxford to Over Wallop on weekends. The puzzle museum became their private paradise.

Roger pulled his car up to James's cottage with Vanessa by his side. James presented Roger with a formidable new rope-and-ring puzzle he'd just acquired. Vanessa woke the next morning to find Roger in the loo, having stayed up all night working on it. They went for long walks in the English countryside, ate simple meals of soup, cheese, and bread, and played with puzzles from one of the finest collections in the world. It was sublime.

The magic and excitement of the relationship came at a steep price for Vanessa: Roger could no longer be her thesis advisor. Her DPhil and the whole of her academic life were doomed from the day he first asked her to dinner, though Roger didn't understand why he couldn't be both her advisor and husband.

University administrators weren't much help in enlightening him. When Roger told the warden of his college that he was planning to marry one of his students who was thirty-four years his junior, the warden said, "Well, it's not as though you're breaking any records."

Roger didn't hound Vanessa out of her degree or her career as a mathematician, the way Lionel had harassed Margaret. Nevertheless, she had no choice but to leave her DPhil unfinished when they married. Roger's work remained unaffected by their relationship.

In 1988, they moved to the village of Thrupp just outside Oxford, buying a tiny, peaceful cottage overlooking the canal. The attic was overrun with mice. Vanessa wanted to set traps, but Roger couldn't bear the thought of killing helpless rodents. He insisted they could live with them. "I had this thing about mice. I couldn't bonk them on the head or poison them or anything like that," he said.

Setting an important precedent in their relationship, Vanessa did not relent and forced a compromise. She temporarily moved back in with her parents while Roger dealt with the infestation. Roger disassembled a standard trap and combined its parts with supplies from the local supermarket. He built a better mousetrap. "Instead of the spring snapping on the mouse and killing it, it controlled a door so that when the mouse got in and got the cheese off it, this little thing would close the door behind the mouse and it would be trapped inside this cage."

He walked each captured mouse out of the village, over the canal and across three fields, to release it in a copse of trees. Vanessa returned to a mouse-free cottage and a better sense of the boundaries she would need to enforce.

In 1989, Margaret was admitted to Comberton Hospital. Roger and Shirley took turns staying with her. Shirley rang Roger from the hospital late one night to say Margaret was dying. Roger drove there as quickly as he could but arrived too late to say goodbye.

In the years since Lionel's death and Margaret's rebirth, Roger had felt closer to her than ever before. As he grieved this new loss, he found comfort in the theory of relativity.

Every moment in a relativistic life exists with equal reality, though we only experience them in sequence. The dead are never really gone, and the yet-to-be-born are already a part of the continuum. Roger's loss was mitigated by the knowledge that the passage of time was a human illusion.

He later discovered Albert Einstein had a similar experience after one of his closest friends died. "Now he has departed from this strange world a little ahead of me. That means nothing. For us believing physicists, the distinction between past, present, and future only has the meaning of an illusion, though a persistent one," Einstein wrote.

Losing his mother, marrying someone so much younger, and approaching age sixty, Roger had every motivation to experience time as an illusion.

The window of the first-floor office where he worked on *The Emperor's New Mind*, his treatise on the nature of consciousness, overlooked narrowboats passing and passersby strolling along the canal. He wrote on an electronic word processor, printing out chapters for Vanessa to read. Vanessa had become an integral part of his work—not only as an editor and sounding board but also as an intellectual companion. They travelled to conferences and symposia together. He valued her opinions and modified his writing and research based on her feedback.

Stephen Hawking happened to be working on a popular science book at the same time as Roger. Their theories had diverged, but they maintained contact based on an uneasy mix of encouragement and competition.

"Every equation you include in your book will halve your sales," Stephen told him over the telephone. But Roger couldn't help himself. He couldn't forsake the details in his efforts to reach a broad audience.

Stephen's book, *A Brief History of Time*, featured just one equation and became one of the best-selling science books of all time. Roger, driven by the optimistic conviction that anyone could and would understand the science, packed *The Emperor's New Mind* with equations and advanced physics from start to finish. He wrote in the introduction,

We shall need to journey through much strange territory—some of seemingly dubious relevance—and through many disparate fields of endeavour. We shall need to examine the structure, foundations, and puzzles of quantum theory, the basic features of both special and general relativity, of black holes, the big bang, and of the second law of thermodynamics, of Maxwell's theory of electromagnetic phenomena, as well as of the basics of Newtonian mechanics. Questions of philosophy and psychology will have a clear role to play when it comes to attempting to understand the nature and function of consciousness. We shall, of course, have to have some glimpse of the actual neurophysiology of the brain, in addition to suggested computer models. We shall need some idea of the status of artificial intelligence. We shall need to know what a Turing machine is, and to understand the meaning of computability, of Gödel's theorem, and of complexity theory. We shall need also to delve into the foundations of mathematics, and even to question the very nature of physical reality.

Roger saw the universe in ways few others could experience. He shared his world with those closest to him: Lionel, Joan, Judith, Ted, Salley, Wolfgang, Tristan, Vanessa, and others. He wanted this book to bring that magic to the masses—complete with the mathematics that made it all real.

Vanessa knew it was risky for Roger to venture so far from his field of expertise, but his reasoning seemed solid enough. She worked to keep him on track and to ensure his arguments made credible sense.

He argued that consciousness researchers didn't have their theories correct yet because quantum physicists didn't have *their* theories correct yet. Consciousness, he argued, emerges from the quantum properties of particles that make up our brains and bodies. Because human understanding is non-computational—as proven by Gödel's incompleteness theorem—some yet-to-be-discovered aspect of quantum physics must also be non-computational.

Roger knew he was venturing far outside the mainstream. But challenging scientific consensus had worked for him more than once.

His worst-case scenario was that the book made no impact whatsoever. "I had no idea what it would do. I thought it would probably disappear without a trace. Then Stephen got Carl Sagan to write a foreword for *A Brief History of Time*. I thought, 'Who do I know? Oh, Martin Gardner!'"

Martin agreed to write a foreword.

"Then I thought, 'Okay probably it won't disappear without a trace.' His foreword would probably pick up a bit of attention. What I was not expecting was the invective there would be from... almost everybody. Sure, I knew the AI people wouldn't like it because I was saying they're not really doing AI. And I knew the religious people wouldn't like it because I wasn't taking a religious point of view. I didn't expect the philosophers would hate it. They mainly hated it not because they disagreed with me, but because they said I was sloppy."[11]

In fact, computer scientists, philosophers, neuroscientists, and other physicists lined up to disagree with him. "Penrose doesn't believe that computers constructed according to presently known physical principles can be intelligent and conjectures that modifying quantum mechanics may be needed to explain intelligence," wrote Stanford computer scientist John McCarthy.[12] "One mistaken intuition behind the... belief that a program can't do mathematics on a human level is the assumption that a machine must necessarily do mathematics within a single axiomatic system with a predefined interpretation." A true-but-unprovable statement in one framework is just an "ordinary theorem" in another, he said, eliminating the need to invoke Gödel.

In the *Times Literary Supplement*, American philosopher Daniel Dennett called the book "the ultimate academic shaggy dog story, a tale whose fascinating digressions outweigh the punch line by a large factor."[13] Dennett questioned why Roger bothered to write this book. Who was he trying to convince, really?

He repeatedly acknowledges that his colleagues, who already understand the difficult materials he is teaching us much better than we ever will, do not yet accept his idiosyncratic vision. But if he can't convince them, pulling out all the stops, what good will it do if he convinces us with a relatively elementary version?...I suspect he has a more subtle strategy in mind. When experts talk to experts, they are careful to err on the side of underexplaining the fundamentals. One risks insulting a fellow expert if one spells out very basic facts. There is really no socially acceptable way for Penrose to sit his colleagues down and lecture to them about their oversimplified and complacent attitudes about fundamentals. So perhaps educated laypeople are only the presumptive audience for this book, hostages to whom he can seem to be addressing his remarks, so that the experts—his real target audience—can listen in, from the side, without risk of embarrassment.

Dennett's review hit home.

There's probably some truth in that. But one shouldn't say that I was simply talking to the experts. I was trying to educate the general public about certain things, not just to do with the AI aspect, but to explain things that I found exciting and interesting about science in a way that was a bit different from most people's way of talking about them. I was certainly talking about science in a more serious way. I wasn't thinking of it as just simply a popular book. I was trying to express viewpoints I hadn't seen expressed.

While I was writing the book, I learned a certain amount about neurophysiology, sort of standard stuff. I came to the conclusion you had to find something in the brain with quantum coherence. It's a huge challenge and I'm not surprised people think it's crazy. It is crazy. But I think it's right. And there are crazy things out there in the world. I mean, quantum mechanics is crazy. Relativity is crazy. What are these things? Even the Earth going round the sun is crazy. It's the natural thing to think that the Earth is stationary and the sun goes around it. You can see it going around. But some of these things which seem crazy at first, we're so used to now that we don't think they're crazy. We know the Earth goes around the sun.

In peer-reviewed journals and public reviews, most authors showed deference to Roger even as they disagreed. "He argues that we lack a fundamentally important insight into physics, without which we will never be able to comprehend the mind. Moreover, he suggests, this insight may be the same one that will be required before we can write a unified theory of everything," wrote Timothy Ferris in the *New York Times*. "This is an astonishing claim, one that the critical reader might be tempted to dismiss out of hand were it broached by a thinker of lesser stature."[14]

"Penrose's book will evoke different responses from different classes of people," wrote geneticist John Maynard Smith. "The AI community will not like it....Biologists are also likely to be critical.... Physicists will like the book. I observe among my physicist friends a conviction that there must be something wrong with biology, because it ignores modern physics. The people who are going to like the book best, however, will probably be those who don't understand it."

Stuart Hameroff, an American anaesthesiologist and professor at the University of Arizona, loved the book. "He had a mechanism for consciousness and nobody else did. Everybody else had kind of handwaving arguments about complexity and emergence....He had a specific mechanism that I didn't understand at the time, and still don't completely understand," said Stuart, who studied tiny structures within neurons known as *microtubules*.[15] As a medical student, he researched their role in causing or preventing cancer. He suspected they might also play a role in information processing. Microtubules exist at the borderline scale between quantum and classical physics—exactly where Roger hoped to discover new physics of quantum coherence, non-computability, and consciousness.

"Stuart writes to me and says, 'Evidently you don't know about these little microtubules.' And he gives me the argument," Roger recalled. "You can talk about crazy things he says, but this was a damn sensible thing."[16]

At their first in-person meeting in the early 1990s, they combined their incomplete and seemingly complementary ideas into a theory of *orchestrated objective reduction*, in which consciousness (and free will) arose not from neural networks but from non-computable quantum information processing in microtubules.

"Stuart's job is to put people in an unconscious state. What do these anaesthetic gases do? Stuart says it's not chemistry. They are acting by some physical process. I've heard him say many times that general anaesthetics do not directly act on nerves. He argued that they act on microtubules. I don't fully understand his point, but it seemed to me a very strong argument."

Roger understood microtubules about as well as Stuart understood non-computable quantum systems. Each saw in the other validation for ideas they wanted to be true. They emboldened each other.

Daniel Dennett never agreed with Roger and never lost respect for him. "Penrose is wrong in a pretty interesting and clear way. I mean this not in a backhanded way at all. If you can make a really clear and tantalizing mistake, that's very useful. Many of the advances of science have come from the correcting of other people's mistakes," he said.[17]

Stuart Hameroff was another story.

"I think Stuart's a nut," Dennett said. "He's very aggressive. He shoots off his mouth. When *The Emperor's New Mind* came out, it was pretty clear Roger knew diddly about neuroscience. But he recognised that, and he realised he should learn something more about the brain. But it turned out he had very bad taste in neuroscientists. I mean, I could name forty or fifty neuroscientists that would have been a better Virgil to lead him into neuroscience than Hameroff."[18]

Discover magazine quoted Stuart's critics calling him "pugnacious," "casual bordering on slovenly," and a "gadfly in the fields of neuroscience and philosophy."[19]

Roger didn't care. "I sometimes felt uncomfortable with collaborations where we would be joint authors. He would go off wildly in some direction into areas I really didn't know about. But I think in this very specific area, he is respected. The trouble is, he doesn't stop himself from shooting off into areas where he sort of loses control of himself," he said.[20]

Their affiliation tarnished Roger's reputation and burnished Stuart's. Steve Volk wrote in *Discover* magazine,

If Hameroff proposed these ideas himself, he might have been ignored, but his co-theorist was Sir Roger Penrose, an esteemed figure in mathematical physics. Their theory, dubbed "orchestrated objective reduction," or Orch-OR, suggests that structures called microtubules, which transport material inside cells, underlie our conscious thinking. But the Penrose-Hameroff model of what you'd call quantum consciousness was a scientific non-starter. Leading experts dismissed the new model outright. Quantum effects, the criticism went, are notoriously difficult to maintain in the lab, requiring ultracold temperatures and shielding to protect against even the mildest interference. Critics said living things are simply too "warm, wet, and noisy" to allow significant quantum effects to persist. What's more, neuroscientists argued, the Penrose-Hameroff model offered no testable hypotheses.[21]

Alongside such public criticism, Roger's friends grew increasingly concerned. Lionel Mason watched with unease to see Roger "aggrandize a lunatic like Hameroff."[22]

Vanessa worried also. "Roger never, ever doubted his own ability to manage these relationships to get what he thought was important out of them," she said.[23] Roger felt his own intellect and reputation would overpower whatever risks these collaborations posed. His confidence distressed Vanessa.

To those who cared about him, he seemed to be working against his own interests. His self-ejection from the mainstream made it harder for people like Lionel Mason who were trying to build on Roger's more viable ideas. "One of Roger's tendencies has always been to regard himself as a maverick who is outside the mainstream and telling them all that they've got it all wrong. To a certain extent I have some sympathy, but he built up a level of prejudice that meant it was very hard for him to accept ideas from the mainstream that were important for us on twistor theory. I regard that process as having held up the subject matter by about a quarter of a century," Mason said.[24]

Mason sometimes found it hard to watch Roger's rejection of mainstream ideas. "You see it in some people: they just want to be mavericks. They want to be the lone outsiders railing against the establishment. And quite often I wanted to say, 'Look, you are the establishment. You have been invited to speak in three different institutions in the last two weeks. So what do you mean?'" he said.[25]

In 1994, Roger published a follow-up book, *Shadows of the Mind*, in which he doubled down on microtubules and orchestrated objective reduction. In the five years between the books, doubt crept in—not about whether he was right but about whether he could convince others. He asked merely to be heard. "Whilst I have learnt not to expect everyone to be persuaded by the kinds of arguments that I shall be presenting," he wrote, "I would suggest, nevertheless, that these arguments do deserve careful and dispassionate consideration."[26]

In the book's opening fable, a girl named Jessica collects plant samples in a cave with her scientist father. When she asks him a question, he dismisses her out of hand, returning to his work. "Jessica hated her daddy when he was in these moods—no she didn't; she always loved her daddy, more than anything or anybody, but she still wished he didn't have moods like that. She knew they were something to do with him being a scientist, but she still didn't understand. She even hoped that she might someday be a scientist herself, but if she did, she'd make sure she never had moods like that."[27] His own childhood spilled unconsciously through the cracks of his writing.

He retold Plato's Allegory of the Cave, arguing that most people—even scientists—remain imprisoned by their own intuitions, requiring a fortunate soul capable of emerging into blinding light to show them the physical world's true nature. This is how Roger saw himself. Others had stripped away pieces of the illusory world, but he had ventured further than any of them into the light. "The hardest thing would be to try to persuade them that any outside world existed at all. All they would know about would be the shadows, and how they moved about and changed from time to time. To them, the complicated wiggling shadows and things on the cave wall would be all that there was to the world. So, part of our task would be to convince people that there actually is an outside world that our theory refers to," he wrote.[28]

He built *The Emperor's New Mind* and *Shadows of the Mind* on metaphors about seeing what others could not. He was the little boy calling out the emperor's nakedness. He was the Platonic visionary emerging into the light.

He used a simple chess puzzle to illustrate how even the best computers of the era had no human-like understanding. A jagged line of white and black pawns traverses the board such that no pawn can advance or capture any other. White's king lurks behind the wall of pawns. Black has a bishop and two rooks, one of which is endangered by a pawn. Any experienced chess player could see that white could play to a draw merely by moving its king back and forth. But Deep Thought, the best chess-playing algorithm at the time, broke the line of pawns to capture a black rook. With its pawn defence destroyed, black easily achieved checkmate.

Roger Penrose designed this chess puzzle to demonstrate the non-computability of human understanding.

"How could such a wonderfully effective chess player make such an obviously stupid move?" Roger wrote. "The answer is that all Deep Thought had been programmed to do, in addition to having been provided with a considerable amount of 'book knowledge,' would be to calculate move after move after move—to some considerable depth—and try to improve its material situation. At no stage can it have had any actual understanding of what a pawn barrier might achieve—nor, indeed, could it ever have any genuine understanding whatsoever of anything at all that it does."[29]

Roger inserted Stuart into the centre of his thesis. "Hameroff and his colleagues have argued for more than decade that microtubules may play roles as *cellular automata*, where complicated signals could be transmitted and processed along the tubes as waves of differing electric polarization states of the tubules....If it is microtubules that control the activity of the brain, then there must be something within the action of microtubules that is different from mere computation."[30]

Roger had burned through most of his supply of intellectual goodwill. Reviews of *Shadows of the Mind* were blunter. Hilary Putnam wrote in the *New York Times*,

> *Shadows of the Mind* will be hailed as a "controversial" book, and it will no doubt sell very well, even though it includes explanations of difficult concepts from quantum mechanics and computational science. And yet this reviewer regards its appearance as a sad episode in our current intellectual life. Roger Penrose is the Rouse Ball Professor of Mathematics at Oxford University and has shared the prestigious Wolf Prize in physics with Stephen Hawking, but he is persuaded by an argument that all experts in mathematical logic have long rejected as fallacious....He mistakenly believes that he has a philosophical disagreement with the logical community, when in fact this is a straightforward case of a mathematical fallacy.[31]

His stubbornness about Gödel's theorem and devotion to Stuart's microtubules became dogmatic. Even fellow physicists tried to intervene. "I get uneasy when people, especially theoretical physicists, talk about consciousness," Stephen Hawking wrote. "His argument seemed to be that consciousness is a mystery and quantum gravity is another mystery so they must be related."[32]

Roger bristled at having his ideas reduced this way. "I mean, my argument is completely the opposite of Stephen Hawking's 'Here's a mystery and here's a mystery.' How does non-computability come out of these physical laws which seem to be entirely computable?"[33]

Roger built a career on seeing things other people missed. He didn't understand why people wouldn't at least entertain the idea that he might be right this time too.

In 1998 at age sixty-seven, Roger stepped down as the Rouse Ball Professor of Mathematics. His body of work would be the envy of any theoretical physicist. He had published hundreds of papers in highly varied research areas. He had transformed modern thinking about black holes, cosmology, and gravity. When he spoke publicly, he packed rooms to overflowing and was greeted with reverence for his science, mathematics, art, and philosophy.

"Let me make an analogy of a ballet dancer," said Ivette Fuentes, an esteemed theoretical physicist at the University of Southampton (and former ballet dancer) who worked with Roger later in his career.

> You can have someone that goes to class all the time and has amazing physical ability and develops an incredible technique. If you want to be really brilliant, you need that technique. If your heart tells you to jump, you can jump and it looks beautiful. But the thing is, you can go and see ballet dancers, and they might have this sharpness and this perfect technique and they're like robots. You don't feel a single emotion. There are other dancers who don't have great technique, but holy moly, the way they stand, the way their presence is—it's like this thing that's coming from inside their soul. It's a creativity you just cannot put into words. What makes a brilliant dancer is the combination of the

two things. I think that's what sets Roger apart: he has the technical ability—his mathematical skills are the best in the world—but he has the other side, which maybe a lot of very technical people are lacking. It's something that comes from another place. The creative part. That's the magic.[34]

He made the most difficult mathematical manoeuvres effortless and delightful and executed graceful leaps of logic beyond what any amount of training alone could explain. His legacy would have been plenty for most people. And still he felt he had more work ahead of him than behind.

As professor emeritus, he had to move to smaller office at the Mathematical Institute, where he continued to supervise graduate students. Vanessa realized they needed more space. "He was retiring, and where were his books to go?" she said.[35]

Brian Aldiss, a science fiction author Roger had recently collaborated with, was selling his home in Boars Hill, a hamlet a few kilometres outside Oxford. "Woodlands" was a spectacular two-storey, seven-bedroom, Edwardian brick mansion surrounded by towering hardwoods. It backed onto the Vale of White Horse, an extensive tract of woods and farmland dotted with clear streams and pleasant trails, perfect for contemplative walks. The massive, high-ceilinged drawing room provided plenty of space for Roger's group meetings with graduate students. He would have his choice of bright, quiet rooms in which to think and write. It was perfect. Unfortunately, by the time Roger and Vanessa knew it was for sale, Brian already had another offer.

Roger left for a conference in the United States, while Vanessa got to work. Brian wanted to sell to Vanessa and Roger rather than to strangers. He agreed to reject the other offer. Vanessa scrambled to buy Woodlands and to sell Canal Cottage. When Roger returned a few days later, Vanessa had acquired Woodlands.

On the day they moved in, Vanessa discovered that Roger's work had arrived in the house before they did. "A person living there before us had left a toilet paper roll in the toilet," Roger said. "Vanessa looked at it and said, 'Have you seen that pattern on the toilet paper? I think that's your tiling.' And it was!"[36]

Kimberly Clark Ltd had reportedly chosen the non-repeating pattern for their quilted toilet paper because tessellating quilting could cause "nesting," in which the quilt lines pile up on each other, creating bulky unevenness in the roll.[37]

Roger's instinct was to take no action. At that time, though, it happened Roger was involved with a company called Pentaplex, which was exploring the idea of turning Roger's creations into commercial puzzles and games. Pentaplex decided to sue Kimberly Clark.

Roger discovered that the company had copied his designs incorrectly. "They were claiming that the reason they used this non-periodic pattern was to avoid the paper sinking into itself. However the pattern they used was actually periodic," he said. "The lawyer asked me to call all kinds of people, including [famed Princeton physics professor] Paul Steinhardt, because Paul had written an article in *Scientific American* that showed a big array of my rhombuses. It became clear that this is what they had copied."

The case was settled out of court, and the consequences were kept confidential. Penrose tiles never again surprised anyone entering the loo.

When Roger was sixty-eight and Vanessa thirty-four, their son Max was born. They both felt lucky to have a family, to be travelling and discovering together, and to have such a wonderful home in which to raise their son. "My goodness how we all loved that house. It was magical. We'd all agree on that," Vanessa recalled.[38]

As wonderful as life could often be, it was really Roger's life that dominated. With Max in the mix, Vanessa balanced the increasingly challenging role of being Roger's sounding board and intellectual partner with parenting, handling family logistics, and keeping the house running, so that Roger was as free as possible to write, lecture, and think. His reputation and community became all-encompassing.

Roger, who had written two books about human understanding, didn't grasp Vanessa's evolving concerns. She was no longer an awed twenty-one-year-old, swept away by his charm. She saw his blind spots, his stubbornness, and the risks he was taking with his reputation more clearly than anyone. Academics and journalists could criticize his books and his affiliation with Stuart with a kind of intellectualized detachment. She saw the potential costs on a more personal level. He sometimes seemed determined to give away his legacy.

Roger hardly noticed her concern or anyone else's. "I don't remember any of my colleagues just sort of slapping me over the wrist and saying, 'Look, you shouldn't get yourself mixed up with this crazy stuff.' I don't remember anybody ever saying that except possibly Vanessa. And I'm not sure whether even she quite said it openly like that," he said.[39]

His friends and family weren't trying to punish him. They were trying to protect him. They offered advice, not admonishments. Roger, though, was more receptive to encouragement from Stuart than cautions from people who cared about him.

17

CYCLES

What if none of this was new? What if the universe was completely unoriginal? What if every mistake, every triumph, every missed opportunity, and every seized one had happened before and would happen again an infinite number of times?

Or what if the cosmic pattern keeps changing, like Penrose tiles? Choices in this timeline might be altered in another. In some alternate universe, Lionel might have been kinder and Margaret freer to express herself. In that timeline, perhaps Roger conquered the universe with Judith at his side; John F. Kennedy survived an assassination attempt, and Roger never went on to invent twistor theory; and Vanessa carried on with her studies and won her own Nobel Prize.

In an infinite universe, anything possible is inevitable. Infinity and endless possibility could arise several ways:

1. In the *many-worlds interpretation* of quantum mechanics,
    new universes are created every time a quantum measure-
    ment is made. Each branching universe differs slightly

from every other, each a separate manifestation of the infinite realm of possibility. Roger rejected many-worlds theory as untestable and absurd—the only universe we can study and understand is the one we live in.

2. Our existing universe could simply be infinitely large. The edge of the visible universe is 46.5 billion light years away. Beyond what we can observe, the cosmos might continue forever. If so, it would inevitably contain infinite copies and variations of Earth. Roger remained agnostic on whether the universe was infinite.

3. Infinite universes could be joined end to end through time. They might all be identical or have all the theme and variation of an infinitely evolving physical reality. Some theories of an *eternal universe* involved cycles of collapse and re-expansion. Roger, though, believed in an endlessly expanding universe that goes through periodic cycles of death and rebirth.

Conformal cyclic cosmology (CCC) was radical. Roger developed CCC slowly, driven by aesthetic instinct and simmering frustration with conventional theories. "I always regarded inflation as a very artificial theory," he said. "The main reason it didn't die at birth is that it was the only thing people could think of to explain what they call the 'scale invariance of the cosmic microwave background [CMB] temperature fluctuations.'"

Inflation, a widely accepted hypothesis that the universe expanded at faster-than-light speeds in its first instant, not only explains why the CMB looks the same at every scale in every direction but also accounts for the observed "flatness" of space. Roger believed it raised more questions than it answered. How did inflation start? Why did it stop? He wanted a cosmic story with no extremely unusual circumstances to get things rolling.

Once more, the "awesome responsibility that the possession of special talents brings with it" impelled him to step up and stop mainstream science from stampeding off in the wrong direction. What was the point of having such a prominent reputation if not to speak truth to power?

He had a growing stack of honorary degrees from universities across the United Kingdom and from Belgium, Greece, Canada, India, and Poland. He was knighted by Queen Elizabeth in 1994. In 2000, the queen appointed him to the British Order of Merit, which rewards military service or excellence in science, art, literature, or culture; it has only twenty-four living members at any time. It is more exclusive than the Nobel Prize or knighthood.

He was accustomed to dining with royalty and flying around the world to accept accolades and pack lecture halls. In October 2003, he gave a public lecture at Princeton University to a sellout crowd of nearly five hundred scientists and fans. He chose this moment to take his stand against ideas he found contemptible. His talk, "Faith, Fashion and Fantasy in the New Physics of the Universe," was a new manifesto about everything right and wrong with modern physics—concerning not only inflation but also string theory, quantum theory, consciousness, and more.

In 2004, he published *The Road to Reality: A Complete Guide to the Laws of the Universe*, his blueprint for repairing everything he saw going wrong with science. He poured all his ideas into the 1,100-page book, including twistor theory, microtubules, cyclic cosmology, and a mercilessly in-depth history, summary, and critique of modern theoretical physics. It took him eight years to write. It was the culmination of all his hopes and frustrations. He railed against string theory and quantized gravity and tried to garner support for twistor theory and gravitized quanta. Among those esoteric debates and extended explanations, he larded the book with his visions of the world-behind-the-world that he still found so beautiful and fascinating.

The *New York Times* called it "a really long history of time." Science writer George Johnson puzzled over why anyone would want to write such a book. Or read it.

Popular wisdom has it that a book everybody bought, and nobody read was Stephen Hawking's whirlwind tour of cosmology, *A*

Brief History of Time. Pictured on the cover, in his wheelchair, the wizened physicist seemed to emanate the gentle wisdom of E.T. But his book, the story goes, was as difficult as if it had been written on Mars.

Far more remarkable, almost begging for explanation, is the success of Hawking's colleague, the Oxford mathematician and physicist Roger Penrose.

. . .

This is very tough going.... The perfect reader for *The Road to Reality*, I fantasized, would be someone comfortable traversing the highest planes of abstract reasoning, yet who had missed some of the most important landmarks of scientific history—a being, perhaps, from another place and time. This is the book alien archaeologists may study for a rigorous, comprehensive view of how the 21st-century inhabitants of the third rock from the sun believed the world worked.

With each year, Roger became more unrelenting. He refused to slow down. He accepted endless media requests and speaking engagements and maintained a punishing travel schedule well into his eighties. An entire world of fans, friends, and colleagues enabled and encouraged him to keep travelling and speaking.

His eyes dimmed, his blood pressure increased, his memory grew less reliable.

Friends and family watched Roger's reputation fade, while Roger worried he wasn't working quickly enough to shore it up. He repeatedly expressed the fear that he'd suffer the same fate as the German climatologist Alfred Wegener. In 1912, Wegener hypothesized that continents move incrementally around the surface of the planet, changing oceanic boundaries and raising up mountain ranges. Critics treated his theory as at least as crazy as anything Roger put in his books—no known physical mechanism could cause such movements. Wegener died years before his continental drift theory was vindicated.

Roger couldn't bear the thought of twistor theory, orchestrated objective reduction, or CCC suffering the same fate. He needed them taken seriously in his own lifetime—if only to ensure work on them continued after he was gone.

"I'm not particularly interested in putting myself forward as a historical figure," he said. "I *am* very concerned that these ideas not get lost."[1]

He returned to the second law of thermodynamics, the principle that entropy will only ever stay the same or increase. A drop of milk is vastly more likely to mix evenly into a cup of coffee than evenly distributed milk is to coalesce into a well-defined blob. Eggs don't spontaneously unscramble, paints uncombine, or decks of cards unshuffle.

If the universe continually gains entropy, it must have started with extreme order. Looking further back through time should reveal everlower entropy. But the very early universe seems to have had extremely *high* entropy. The roiling baryon-photon fog of the first 379,000 years had no discernible pattern or structure. Before a hot sun could burn in a cold, dark sky, before there were massive structures like galaxies and galaxy clusters, and before the intricate unravelling of cosmic order began, the early universe must have had a hidden structure, an untapped reserve of low entropy that could offset the superheated chaos. Where was it hiding?

Roger wrote a paper in 1979 with an answer involving the Weyl curvature hypothesis. The early universe's gravitational field hid more than enough low entropy to compensate for the way randomness was maximized in every other way. This special gravitational state created an overall curvature of space-time (known as the *Weyl curvature*) close to zero, which created the initial conditions for our current homogenous, self-similar universe.[2] Gravity created the stars, galaxies, and other structures, hot spots and empty regions, and all the objects astronomers study today. Entropy only increased, thanks to that initial low-entropy gravitational field. "The Weyl curvature is zero when all the degrees of freedom are suppressed. People don't seem to emphasize that, but it seems like the key thing to me," Roger said.[3]

Roger knew other cosmologists weren't going to like his explanation for how gravity arrived at that highly structured state. "There is a profound oddness underlying the Second Law of Thermodynamics and the very nature of the Big Bang. In relation to this, I am putting forward a body of speculation of my own, which brings together many strands of different aspects of the universe we know," he wrote. "The scheme that I am now arguing for here is indeed unorthodox, yet it is based on geometrical and physical ideas which are very soundly based."[4]

He asked physicists to let go of the very idea that the universe began with the Big Bang. The low-entropy universe didn't just burst into existence 13.7 billion years ago. The story began much earlier. To prove it, he returned to one of his favourite areas of mathematics: conformal geometry.

He let his mind drift once more to the distant future. The universe as a whole was expanding, but gravity would cause denser regions within that expanse—galaxies and galaxy clusters—to collapse. The incessant pull turned old galaxies into new quasars. Ravenous black holes absorbed billions or trillions of solar masses' worth of material. Stars winked out of existence. Behind ever-growing event horizons, information was destroyed forever. The universe grew so cold and empty, hyper-massive black holes began to boil away with Hawking radiation. It might take a googol years before the largest black holes completely evaporated, leaving the universe sparsely populated with high-entropy radiation—an endless expanse of massless photons.

The thing that started me off was really thinking about the photons and how boring the remote future was going to be. All the black holes gone. Just these photons drifting forever and ever and ever. This just struck me as incredibly boring. This is a very emotional argument, if you like. But then I thought,

"Who's going to be bored in this universe?" Well, mainly photons. But it's very hard to bore a photon. Not because a photon presumably has no actual experiences. That was not the point. The point is that time for a photon is always zero. You just look at the way time dilation works: the time with respect to that photon from creation right through to infinity is zero.

So where do these photons go? They don't want to be cut off from existence. Is there something on the other side? What could it be? That was the point where I thought, "Hey, could it be a stretched out big bang?"[5]

Massless particles do not experience time. They also don't experience space. In a universe full only of photons, distance and scale cease to exist.

Roger let his mind drift *backwards* 13.7 billion years to the cosmic pinpoint that supposedly started it all, when the universe was trillions of degrees too hot for matter to form. It too was filled with massless particles. It too had no clocks and no rulers. The beginning and the end were both conformally invariant. At each terminus, scale disappeared. Viewed through the lens of conformal geometry, each was indistinguishable from the other.

He recalled M. C. Escher's iconic lithograph "Print Gallery." A boy stands in a gallery looking at a print of a cityscape. The lines curve ever outwards, the gallery print expanding into an actual city. The city contains a gallery, and the gallery contains the boy looking at the print of a cityscape. The image expands endlessly and cyclically, passing through the same starting point.

Could this be the shape of the universe? If he could look far enough into the future, would he see the back of his own head? Would he see another universe entirely? Had he found a clever mathematical trick or a physical truth? The world-behind-the-world, the place where curves and angles have greater reality than even time and space, the realm Roger spent a lifetime escaping to, might actually exist at the beginning and the end. These two distant extremes might be one and the same.

M. C. Escher's "Print Gallery."

He hovered where the birth and the death of the universe bled into one another. He travelled further back and further forward to other scale-invariant transitions. An endless cycle of heat death and fiery rebirth. Non-stop expansion. Entropy would always increase, but every so often the perspective would change—a conformal "zoom out to infinity"—such that the large became tiny, cold became hot, old became new, and the cycle began again.

Conformal cyclic cosmology was a difficult sell. To make it work, every last massive particle—all the electrons, protons, and neutrinos, plus all the dark matter—had to be eliminated from the universe. Not all of it would fall into black holes. He required some other physical mechanism to cause rogue particles to "fade away."

He also struggled with the question of falsifiability. Would it be any more possible to test the existence of previous universes than parallel universes? What happened to time's arrow during the transition from one aeon to the next? How could a new universe begin *after* time had been eliminated from reality?

As he drifted through the transition from a dying universe to a cosmos reborn, these challenges all seemed manageable. The story was too powerful, too beautiful, too elegant to be wrong.

CCC eliminated the need for inflation and Stephen Hawking's cosmological singularity. The expansion required to homogenize and flatten the universe no longer had to happen in a tiny fraction of a second. Gravity naturally emerged into a new aeon in an extreme low-entropy state—resolving Roger's questions about the second law of thermodynamics. As a new universe spilled out quarks and gluons, cosmic radiation, stars and galaxies, hot spots and cool spots, rivers and volcanoes, entropy never stopped increasing. The second law could carry on for eternity.

In 2010, he published his fourth popular science book, *Cycles of Time: An Extraordinary New View of the Universe*. He continued to highlight his disagreements with Hawking.

> Hawking himself has been one of the strongest proponents of the viewpoint that information is indeed lost in black holes. Yet, at the 17th International Conference on General Relativity and Gravitation, held in Dublin in 2004, Hawking announced that he had changed his mind and, publicly forfeiting a bet that he (and Kip Thorne) had made with John Preskill, argued that he had been mistaken and that he now believed that the information must in fact all be retrieved externally to the hole. It is certainly my personal opinion that Hawking should have stuck to his guns, and that his earlier viewpoint was far closer to the truth![6]

Information loss was crucial to CCC. "I am thus asking the reader to accept information loss in black holes—and the consequent violation of unitarity—as not only plausible, but a necessary *reality*, in the situations under consideration. What does 'information loss' at the singularity actually mean? A better way of describing this is as a *loss of degrees of freedom*, so that some parameters describing the phase space have disappeared, and the phase space has actually become *smaller* than it was before."

The destruction of information in a singularity allowed the universe to begin again.

People weren't sure what to do with CCC. Reviewers called *Cycles of Time* "heretical," "radical," "speculative," "unorthodox," "not for the fainthearted," "abstruse," "an intellectual thrill ride," "controversial," and "a quite extraordinary hypothesis."

Peter Woit, a Columbia University mathematical physicist wrote a sympathetic review of *Cycles of Time* for the *Wall Street Journal*. He quickly followed up with a blog post reassuring his colleagues that he knew Roger was dead wrong.

> I should make it clear that I'm not at all convinced by what Penrose is proposing. Attempts to get a Big Bang in our future as well as our past generally strike me as motivated by a very human desire to see in the global structure of the universe the same cyclic pattern of death and rebirth that govern human existence. To me though, deeper understanding of the universe leads to unexpected structures, fascinating precisely because of how alien they are to human concerns and experience. Just because we might find a cold, empty universe an unappealing future doesn't mean that that's not where things are headed.[7]

Roger did find the prospect of heat death boring and depressing. CCC appealed to him first aesthetically; only later did he become convinced it was true. In academic circles CCC was met with confusion, discomfort, and an uneasy sense that its author had "gone off the rails."[8]

Roger dug in. He told other physicists exactly where to go to look for evidence for CCC. At the end of an aeon, conformal geometry compressed the unimaginably slow evaporation of a black hole into a near-instantaneous burst of energy. The blast would be powerful enough to pierce the transition between aeons, reverberate through opaque plasma for 379,000 years, and manifest as detectable patterns in the CMB. "The supermassive black hole radiation comes bursting through. What does it do? All the Hawking radiation stretches out to a region which looks like about four degrees in the sky...about eight times the moon's diameter," Roger said.[9] Such spots wouldn't be warmer or cooler than the rest of the CMB but would have less variation overall in energy levels.

In a 2018 paper, published just five months after Stephen Hawking died, he suggested these spots be known as *Hawking points*, ostensibly in recognition of Stephen's contribution to the theory. Were Stephen Hawking alive, though, he would have wanted nothing to do with CCC. The name struck some scientists as more opportunistic than collegial. "He called them Hawking Points, which I found kind of distasteful because Stephen Hawking had just died. If it was done for publicity purposes, Roger Penrose doesn't need publicity," said Douglas Scott, a professor at the University of British Columbia.

Scott was one of many physicists who accepted Roger's challenge to test CCC, more out of respect for the man than the idea. "It was only a prediction because Roger said it was a prediction. Nobody else understood why there would be low-variance circles in the sky. And there's no chain of reasoning that says whether or not you detect these things rules out or rules in CCC. But he said it was a prediction, so we went off and tested it," Scott said.

His team's analysis turned up no evidence of patterns in the CMB that matched Roger's description. By the time they published their response, Roger had already posited another prediction. At a lecture at Princeton celebrating fifty years since the discovery of the CMB, he said there should be concentric rings of low-variance energy in the CMB that were also remnants of the previous aeon.

Scott was in the audience, and once again gave Roger the benefit of the doubt. "I was in two minds of wondering whether to say anything or just keep my mouth shut. I couldn't stop myself. I said, 'You know, if you really think there are multiple rings, we will look for them,'" he said.[10] He put an undergraduate student on the project, who did, in fact, find ring-like patterns in the radiation map but could not find any reason to connect them to CCC. In anything as complex and detailed as the maps of the CMB, such patterns could easily arise randomly.

Princeton physicist David Spergel fared similarly. "Roger reached out to me on the CCC stuff. I was leading the analysis of the [Wilkinson Microwave Anisotropy Probe] data at the time, and he asked me whether I would collaborate on it. I encouraged Amir Hajian, one of the postdocs on my team, to work with him. Amir didn't find the pattern that Roger was looking for. He didn't see what Roger wanted. Roger refused to publish. He didn't want to hear that it wasn't there."[11]

There was no surprise and no joy among scientists refuting Roger's theory. He was liked and respected, which made it all the worse. "I have seen a phenomenon in many brilliant scientists who were able to accomplish a lot by not listening to people and going their own original path. But as they get older, they should be listening more. Instead, they listen less," Spergel said. "You run out of people telling you when you're being an idiot because there's nobody above you anymore in the scientific food chain. When you're young there's bunch of your colleagues who are equally smart and they say, 'Hold on a minute, what the hell was that?'"

As Roger increasingly ignored his critics, he also stopped listening to Vanessa. He no longer cared if his ideas made no sense to her, and she all but gave up on hearing him out. He became hyper-focused on finding anyone who would give his ideas a sympathetic hearing.

"I had this interview when I was in the States with this chap called Joe Rogan. Vanessa said, 'Oh you got involved with him, did you? Well he's all mixed up with these crazy, wild things.' I said I don't have to agree with his views. People have completely different views, and that's not important. Vanessa got very upset with that, and thought I shouldn't have anything to do with him."

Vanessa was more than upset. The Rogan interview revealed a major rupture in their relationship and challenged her basic belief that she and Roger still had any shared values.

Joe Rogan, an immensely popular and influential podcaster, has a history of using racist, misogynist, anti-feminist, fat-phobic, homophobic, transphobic language and of providing a high-profile platform to pseudoscientists, conspiracy theorists, and other perpetrators of misinformation.[12] Rogan mixes credible scientists in with crackpots, making it difficult for people to know which is which.

Roger wasn't interested in the problematic elements of Rogan's show. Rogan was willing to listen, so Roger was ready to talk. Despite responses from people like Douglas Scott and David Spergel, he was convinced the scientific community was ignoring him. Rogan offered him a chance to be heard. He said,

> There's something rather nice about being wrong. It's a rather strange thing to say. If you find out you were wrong about something, it can open things up. You think, "I've been going on the wrong track. *That's* why I was stuck." That has happened to me. What I don't like is what's happening with my cosmology stuff. How much attention does it get? Zero. Now *that* is spooky. If there was something wrong with it, they should tell you. If there's not something wrong with it, they should say, "Gosh that's a good idea." I don't find anything. I just find zero. And I find that extremely frustrating.

In addition to academic responses, CCC was covered widely in popular science media, as well as by PBS, the BBC, the *New York Times*, multiple high-profile YouTubers and bloggers, and major newspapers around the world. Nothing could shake Roger's sense, though, that he was being ignored.

That sense might have been exacerbated by one of his other major ideas getting lost in the shuffle of a Nobel Prize–winning discovery.

Back in 1982, a few years after Roger published his five-axis, aperiodic Penrose tiles, an Israeli crystallographer named Daniel Shechtman discovered a metallic alloy whose atoms were organized unlike anything ever before observed. These structures, which

Shechtman called *quasicrystals*, shared the pentagonal geometry of Penrose tiles—an atomic arrangement previously dismissed as impossible. It was as though Roger had invented a fantastic species of animal, and zoologists had then found that very creature living in the wild.

In 2011, Shechtman was the sole recipient of the Nobel Prize in Chemistry for his discovery of quasicrystals. Roger's work was not recognized. "I thought it was outrageous the way that the chemists wrote him out of the story in the citation for the Nobel Prize," said Lionel Mason, Penrose's longtime colleague at the Oxford Mathematical Institute. "Of course, the problem there was that he published his quasicrystals in a recreational mathematics journal."[13]

Shechtman described Roger's influence on his quasicrystal discovery similarly to how Roger described Johannes Kepler's back-of-the-mind effect on his development of the kite and dart tiles. "I was familiar with Penrose tiles, of course," Shechtman said. But he wasn't consciously thinking of them as he did his research. "I didn't know what it was. I repeated my experiments time and again over the next months. At the end of my sabbatical, I knew exactly what it was not, but I still didn't know what it was."[14]

From Kepler to Penrose to Shechtman, pentagons kept reappearing and revealing new secrets. In this case, Roger was satisfied to let the shapes have a life of their own. He consulted on many art and architecture projects related to Penrose tiles but never pursued prizes or commissions. He did enjoy his growing celebrity. In North America, across Europe, in India and Australia, crowds lined up for hours for a chance to hear him speak. Fans asked for autographs, and he appeared regularly on the BBC and other major media. He felt very well rewarded.

Only with his newer, less adulated ideas did he feel shortchanged.

18

FANTASY

Princeton University Press approached Roger to write a new book based on the lectures he'd given there in 2003. "*Fashion, Faith, and Fantasy* came about because of those three lectures. They kept wanting me to change the title, but I didn't want to because it was exactly what I meant," he said.[1]

Resentful of the perceived silence that greeted his theories of consciousness, cosmology, and computability, he was ready for a dust-up. String theory remained inexplicably fashionable. Quantum mechanics elicited a dogmatic faith in unitarity and the Schrödinger equation. Inflation and much else of contemporary cosmology were pure fantasy.

The Guardian newspaper called *Fashion, Faith, and Fantasy in the New Physics of the Universe* "an explosive study from an eminent refusenik." "It seems from *Faith, Fashion and Fantasy* [*sic*] that Penrose has not felt comfortable with any of the radical new ideas in fundamental physics that have been set out in the past 40 years," science journalist Graham Farmelo wrote in the paper's review.[2]

Roger felt a responsibility to call out the smugness and complacency of modern physics. "If a proposed scientific theory can be revealed as being too much influenced by the enslavement of fashion, by the unquestioning following of an experimentally unsupported faith, or by the romantic temptations of fantasy, then it is our duty to point out such influences, and to steer away any who might, perhaps unwittingly, be subject to influences of this kind," he wrote.[3]

Fashion, Faith, and Fantasy didn't garner the same attention as his previous books. He was no longer a renowned physicist stepping out of line to present daring alternative theories. He was a known maverick who had departed long ago from the mainstream. "In standing outside the fray and criticising the central dogmas of fundamental physics, Penrose is playing the role of Einstein, who forced quantum theorists to defend and hone their ideas, and Sir Fred Hoyle, who persistently challenged Big Bang theorists to sharpen their ideas. This is an extremely important role, and long may Penrose fulfil it," Marcus Chown wrote in the *Times Higher Education*, echoing Daniel Dennett's sentiment that Roger was wrong in an interesting way.

He continued working with Stuart Hameroff on microtubules and cultivated new collaborators to hunt for evidence of conformal cyclic cosmology.

He was so preoccupied with finding validation for his ideas that he continued to ignore concerns from his allies.

"There was an occasion I had these pictures of the centres of these triple rings selected by the low variance of their temperatures. I tried to show Vanessa and she deliberately took no interest in it. I took this as saying that she thought I was barking up the wrong tree and that my ideas were crackpot and so therefore she wasn't interested in them. I wouldn't say it was an argument exactly, but it was a disturbance," Roger said.

He wanted support rather than the honest feedback Vanessa did her best to give him. This was his Alfred Wegener moment. Why couldn't she see it? "If you were Mrs. Wegener, would you have said, 'Look, you are being arrogant now' and all that? Shouldn't you have been supportive of him and said, 'Yes, dear. I think your idea is very good' or something like that?"

Vanessa had heard this argument many times before. Roger frequently invoked the continental drift theory, though it bore no relation to CCC. She continued to look after the family's practical needs. She kept their lives running but could not reach him on his scientific work. Roger was no longer willing to listen to any counterarguments.

In the face of difficult personal interactions, old instincts re-emerged. Roger withdrew, fading away from Vanessa and Max, disappearing into the CMB.

In 2012, Ivette Fuentes, who was then based at the University of Nottingham, sat in Heathrow's waiting area, going over slides for the presentation she was on her way to deliver at the Ninth Vienna Central European Seminar on Dark Matter, Dark Energy, Black Holes, and Quantum Aspects of the Universe. Ivette, a Mexican-born quantum physicist, studied relativistic quantum mechanics. Like many younger scholars, she mixed theoretical research with her own experiments and pursued both knowledge creation and applied science.

As she waited for the boarding call, something colourful across the room caught her eye. Someone held up a drawing with bright oranges, blues, and greens. It took her a moment to place the image. "And then I thought, 'I recognise that. I know that picture. Where have I seen it? Oh, that's Roger Penrose's slide!' And then I looked up, and I saw it was Roger."

They had never met but she knew his reputation well enough to be intimidated. Which meant she had to talk to him. "When I fear something, I go, and I do it. That's a commitment I have with myself," she said.[4]

She sat down in an empty seat next to him.

"You don't know me. My name is Ivette Fuentes and I know you from your work. And I think we're going to the same conference in Vienna."

They chatted for a few minutes about the slide, "Schrödinger's Little Mermaid," which was part of the talk he'd prepared for the conference. From that first conversation, Roger found Ivette's views on physics extraordinarily sensible. Could she be another Ivor, Ted, or Wolfgang, he wondered? Or was there any possibility she could be another Judith?

They were called to board. Roger went off to first class, Ivette to economy. She saw him again at the Vienna airport. He was surrounded by colleagues and fans, and she didn't interrupt him. She was surprised to see him again the next day, listening attentively in the front row at her talk, which focused on the potential to use relativistic motion to manipulate quantum information.

Roger showed great interest in her work, which, like his, was somewhat outside the mainstream. Their situations, though, were not really that similar. Roger cultivated his outsider status. As a racialised woman working in elite physics, Ivette had no choice about hers.[5] Regardless, their intellectual bond grew into both a professional relationship and a friendship.

Roger became an enthusiastic participant in Stuart's annual Science of Consciousness conference in Arizona. At these conferences, his work searching for a physical explanation for human understanding mixed with discussions of "neurospirituality," "conscientiology," and other areas of quasi-philosophy and experimentation very far from his own approach.[6]

In late 2016, Stuart introduced him to James Tagg, an American entrepreneur who was deeply into Roger's ideas about consciousness. James proposed he found a Penrose Institute to further this research. Roger agreed to lend his name to the project, whose mandate, funding, and governance structure were very loose.

Despite a lack of affiliation with an academic organization or understanding of academic research institutes, Tagg was very keen to focus on Roger's consciousness research and to link it to even more outlandish ideas. Roger ceded control over how his name and ideas were used. The people concerned about Roger's alliances with Stuart Hameroff and Vahe Gurzadyan were deeply disturbed by his apparent willingness to let James Tagg take over his legacy.

"I think the main danger there was that he would put family funds into it. I think that must have been something that frightened Vanessa," said Lionel Mason. "I nominated him for a Breakthrough Prize, and when I heard about the Penrose Institute, I thought, 'Oh god, if he gets the prize, he's going to dump it all into this Penrose Institute.'"[7]

In reality, Roger had no family funds to invest. Nevertheless, the institute was a breaking point for Vanessa. He was not only blind to the damage he was inviting to his reputation but deaf to anything she had to say about it. His disregard for the effects of his decisions on her and Max was a painful indicator of just how indifferent he had become.

Roger tried to broaden the institute to support all his research interests rather than just cater to Stuart and James's focus on human consciousness. He asked Ivette, whose work related to his theories on quantum physics and cosmology, to join the project. "It was January 2017. I got an email from Roger saying he would like me to be part of this founders' group, and especially take care of the physics part. It was very cool. It was very exciting," she said.

Roger and Ivette brainstormed candidates for the new institute's scientific advisory board. They left the fundraising to James and tried to set a scientific agenda that might save Roger from Alfred Wegener's fate.

The institute began to unravel before anyone made major investments. Members of the board, including Roger's old friend Lee Smolin, saw the institute was sputtering and resigned their positions. The Penrose Institute never got further than the creation of a website.

James and Stuart didn't let go. At the 2017 Science of Consciousness Conference, Stuart's recklessness and Roger's haplessness nearly destroyed his reputation.

Stuart's wife approached Roger at the conference. "Someone wants to chat with you. Would you mind?" she asked. Roger agreed without asking whom he would be meeting.

I walked over and sat in this chair. Some of these mystical people were there at the conference, and this chair was supposed to make you feel more grounded or more relaxed. The person who I was talking to was Jeffrey Epstein. I was sitting there talking to Jeffrey Epstein and he was asking me about this institute. Now I didn't really know anything about him. He said he had these parties. He was wondering whether I'd be interested to go to New York. He could invite Woody Allen to this party. I was smelling a rat a little bit and said I had no plans to visit New York any time soon.[8]

In 2017, Epstein was still two years away from making national news for trafficking minors. But he had been convicted in 2008 by a Florida state court for procuring a child for prostitution and for soliciting a prostitute. He also was a wealthy, well-connected philanthropist who had a reputation for giving no-strings-attached research grants to people he personally viewed as geniuses.[9]

Roger, Stuart, and James met not long afterward to discuss whether they could get money out of Epstein to fund Ivette's gravitational wave detector experiments and finally get the Penrose Institute off the ground. They floated the idea past her.

"They said there was someone who had offered money for my project," Ivette said. "They didn't say the name, but if they had I would not have recognised it because it was 2017. They just said, 'There is someone who wants to offer money. But there is a problem attached to it, which is it's someone who has been convicted of sex offences.' I didn't even want to know how serious the crimes were. I just immediately said no. And there was no more talk about it. No more discussion."[10]

Roger and Ivette spoke frequently by telephone, sharing ideas about gravitational waves, quantum collapse, BECs, CCC, the CMB, twistors, and space-time. Unwittingly, Ivette was filling a gap in Roger's emotional life simply by being enthusiastic about his work. Roger was only partially aware of it himself, until he slipped up on a call late one night and uttered the word *muse*.

Ivette had built a stellar career in physics, contending with systemic sexism and racism while publishing, lecturing, searching for financing, and fighting for her own scientific agenda. She was focused on her ideas, not on inspiring an octogenarian Oxford professor. She was furious.

The fallout lasted for months and nearly ended their friendship and their professional relationship. "I got very angry at him, and I told him, Don't you ever, ever talk like that about me," Ivette said. "I was thinking that it would not interfere or bother me if he could keep it to himself. But at some point, he mentioned again this muse thing, and I felt he started to cross that line. I told him, 'I don't want to talk to you anymore. It's not just about you not calling me your muse. I don't want to *be* your muse.'"[11]

Roger could imagine what the end of the universe looked like to a photon, how a four-dimensional creature would appear to a human being, and what an astronaut who fell into a black hole would experience. But he could not wrap his head around the experiences of a woman working in theoretical physics. He couldn't see how insulting it was to relegate a successful physicist to the role of muse. He could not grasp the potential damage to her career should there be even the appearance that he, a hugely influential figure, supported her work for any reason other than her academic credentials.

He pleaded with her not to "cut him out." She missed their conversations but saw no other option. "It bothered me how people would perceive something like this. It also bothered me that he was perceiving me in a different way. It was very messy. It was very, very messy," she said.[12]

Weeks passed without contact. The Penrose Institute was on hold. Roger buried himself in work. Though he didn't completely understand how things had gone so wrong, he had some dim recognition that he needed to take responsibility.

"He wrote me a really sweet letter and apologised," Ivette said. It was just enough. With boundaries more clearly drawn, they gingerly moved back toward friendship and collaboration.

They discussed creating a revamped Penrose Institute, without James and with better planning and governance. Ivette had moved to the University of Southampton. The university administration indicated willingness to devote organizational and fundraising resources. Stuart remained involved as an advisor.

For Roger and Vanessa, the damage was too great. There was no tolerable version of Roger's involvement with the institute or the people connected to it. Vanessa needed space. She began looking for a flat that would allow her some breathing room. At the time she wasn't envisioning them living separately but seeking a place she could have to herself on weekends.

She discovered a house and a flat for sale across the road from one another on the outskirts of Oxford. She suggested they sell Woodlands and buy these two places so they could effectively "live together separately."

Roger balked. He started looking for places closer to the Oxford campus and found a flat in walking distance from the Mathematical Institute. The sale of Woodlands allowed them to buy all three properties.

Instead of living near his family, Roger chose to be closer to his work. Alone again, he was already lost in research before he'd unpacked.

19

MORE TIME TO THINK

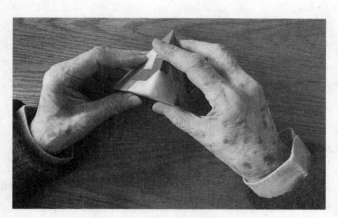

Roger Penrose holds a copy of his tetrahedron puzzle.

On the afternoon of October 5, 2018, I sat at a crumb-covered table in Roger Penrose's flat, about to conduct the first interview for this book.

Roger moved around his kitchen preparing tea and biscuits. He left the lights off. Macular degeneration had left him nearly blind. He navigated mostly by touch and memory. Grey autumn light filtered in through the windows, which overlooked a canal where narrow-boats and swans passed unhurriedly. Stacks of moving boxes filled various corners of his flat, and framed prints and paintings leaned against walls, waiting to be hung.

He instinctively picked up a copy of the tetrahedral puzzle he had designed during his undergraduate years. He slid out a piece that locked several others in place, and six irregularly shaped, multicoloured chunks of plastic clattered onto the table.

He shuffled the pieces around as we talked, feeling out small bumps and indentations that provided clues about how they fit back together. Meanwhile, he wandered from story to story, subject to subject, attempting to assemble the key fragments of his disjointed life. He drifted between cosmology, artificial intelligence, crystallography, black hole physics, and chess puzzles. His talked about higher-dimensional Platonic solids, alternatives to the standard twelve-note octave, the death of his mother, and the influence of Dennis Sciama. Every new tangent left a previous thought unfinished. When he got more comfortable, he disclosed that he and Vanessa had recently separated. Alone for the first time in more than thirty years, he insisted solitude just gave him more time for research.

As he spoke, his hands appeared to be running on a separate system. They kept working the pieces of the puzzle, turning and interlocking, until, with no flourish whatsoever, he slid the final three pieces simultaneously into place. He once again held a smooth tetrahedron, about four inches high. With barely an acknowledgement, he broke the pyramid apart and started again.

"It's all complicated, and it's all connected," he said.

He gave me a tour of the geometric puzzles, games, and oddities he'd amassed through a lifetime of mathematical play: plastic gears and wooden tiles, multicoloured resin cubes that unfolded into jagged shapes, a Klein bottle, a collection of tiny cardboard polyhedra he'd made as a child. There was no sign of his many awards, medals, and plaques, which were stowed out of sight. These toys were the badges of pride and delight he chose to display.

In the years that followed, Roger and I spoke for hundreds of hours by phone, videoconference, and in person. I met his siblings, sons, colleagues, and former spouses. I read his published and unpublished writing. Each interview, notebook page, and old fax added a new piece to the picture. The process by which a human being emerges from their constituent parts—be they quarks and electrons or stories and memories—works differently than a jigsaw. The moment never comes when the last piece clicks into place and the puzzle is complete.

On a snowy day in February 2019, Roger and I walked from his flat to the Ashmolean Museum where Roger's collection of M. C. Escher prints was on long-term loan. I reached out to help steady him on a steep hill where broken ice covered an already treacherous stretch of pavement. He brushed me back.

"If I fall, you can help me up," he bristled, jamming his walking stick into the ice and carrying on.

Later on the same walk, he paused in the chilly air.

"I'm going to tell you something I've never told anyone. It's a theory I think about. I don't think it's true. But I think there's something to it. What if people alive today are visitors from the future who bought tickets to come back and have a twenty-first century experience?"

He absolutely did not believe consciousnesses could jump through time or from body to body. He just liked thinking about the idea. It helped him make sense of his own experiences.

Imagining himself as a ticket-holding consciousness from the future reinforced his self-image as a spectator to, rather than the architect of, his own story. He had trouble taking credit (and responsibility) for his extraordinary life: he hadn't designed the laws of physics that governed how his future unfolded.

I asked him more than once how he had evolved from a relatively obscure pure mathematician to become Roger Penrose™, famed creator of impossible objects, tamer of infinities, and storyteller of the universe.

"It would have been stranger had I become anyone else."

I asked him if he considered himself a generally happy, unhappy, or neutral person. With long pauses, he said, "Listen to the opening of Bach's St Matthew Passion. It's happy and sad at the same time."[1]

We walked up and down Oxford streets, visiting the office he still kept at the Mathematical Institute, various flats where he'd once lived, the two campus areas paved in Penrose tiles, the bakery that made his favourite seed bread, and muddy, wooded paths along the city's waterways where he took his daily walks. He showed me the "Nick Woodhouse shortcut" he'd named for another Oxford maths professor, which entailed squeezing through a gap between a fence and a building to cut off about two metres of the walk across campus.

He mentioned some letters he had written to a young mathematician in the 1970s. He thought they might provide insight into his thinking at the time he was developing twistor theory, Penrose tiles, and other major ideas. He and Judith Daniels had never been a couple, he said, though she had been a significant confidante.

When Judith died of lung cancer in 2005, her sister Helen discovered Roger's letters among her possessions and returned them to him without opening them herself.

Roger was married by then to Vanessa. Although the letters were almost as old as his wife, Roger thought they might make her feel jealous or insecure. He wanted them out of the house before she stumbled upon them. He passed the parcel, still unopened, to his administrative assistant at Oxford University, Fiona Martin, who stored them in her attic.

In April 2019, Roger was invited to speak at the Perimeter Institute for Theoretical Physics in Waterloo, Ontario—a city a little more than an hour's drive from my home in Toronto. He retrieved the letters from Fiona and brought them with him to Canada. I met him at Pearson Airport to drive him to Waterloo. His flight was delayed, and we didn't get in until 2:00 a.m. Before we parted company for the night, he produced the packet, which filled half his suitcase.

I left the bundle unopened on a coffee table. We'd agreed to meet at 8:00 a.m. to see what was inside. He knocked on my door at 7:00.

"Shall we open that package?"

It contained more than two hundred letters, totalling thousands of handwritten pages—much more than Roger had let on.

Even with the magnifying glass embedded in the right lens of his customized glasses, he had to hold a paper a fraction of an inch from his face and move it back and forth to decipher text. We agreed I would read out loud to him instead.

I chose a fat envelope at random from 1974. Roger's boyish, readable handwriting filled page after page, with afterthoughts and postscripts scrawled in the margins.

It only took a few paragraphs to realize that he had been in love with her and she had not felt the same way. I was unprepared for this turn, but not nearly as unprepared as Roger. Behind the thick lenses of his glasses, his eyes were sad and wild as I read his own words back to him.

> I can see mechanisms at work in myself which are emotional as much as physical.... There is a form of missing you, which consists of a fear that you may be forgetting me, and of jealousy that you may be involving yourself more deeply with someone else. The other negative feeling, which I have observed within myself toward you more than ever before, is a certain anger.... I don't completely understand it myself, but somehow over the past several weeks, my feelings toward you have made contact with something deep inside myself. Perhaps it has to do with a feeling of resentment towards Margaret that she had deprived me of one kind of love owing to her own inhibitions, and had substituted another.

We paused.

Roger's words, unseen and forgotten for forty-five years, sent him tumbling into a freefall of long-buried confusion and pain. He felt his way around the edges of these words, piecing together what he had been feeling.

"The thing was my mother was very non-demonstrative in her love for us. She would do anything for us. But there wasn't this sort of warm hugging love about her," he said. "I once had this conversation. It was at the dinner table with me, my parents, and my brothers. And the topic came up about dreams. And I said, 'I sometimes dream about being terribly angry about something, and at a very specific person. And it's very strange.' I said I wasn't going to say who it was. Margaret said, 'I know who it was.' It was her. She knew."

Roger's voice broke as he spoke these words. Tears rolled down his face. Like one of his puzzles, he fell to pieces. He left the room.

Sometime later, he collected himself and rejoined me. We carried on with the letters. In them, he wrote about feeling that Joan had tricked him into marrying her. He disclosed his shame at not being able to trust in his relationship with Judith. He lamented his "hopeless" love for her. He was often awash in misery but also observing that misery from a distance, trying to work out what external mechanisms had led up to such an unhappy situation. At every turn, he sought ways to solve the puzzle so he could continue talking to her about physics. He was sad when she rejected him romantically but cared much more about their scientific connection.

This was why he had wanted me to read these letters.

Roger had career successes before and after Judith. But the connection he felt to her, reawakened through his old letters, overpowered him. He saw her as his "muse." She opened doors in his mind, elicited new ideas and theories, and made his work feel meaningful. When their paths separated, he never felt the same about his ideas again.

He wept for the loss of his source of scientific inspiration. He wept for his anger at his mother, his estrangement from Joan and their three sons, Salley's departure from his life, and his separation from Vanessa. We had spoken about these things before, but Roger had always intellectualized them—they were just more brainteasers for him to tinker with. Judith's letters revealed the pain underlying his bafflement.

Even with this loss laid bare, he still did not see his own role in how he had suffered and the price the people around him had paid so he could be a "lone genius." Life had happened to him—he discovered rather than created it.

After several hours of reading, we sat in silence, contemplating the hundreds of letters that remained. We both had business at the Perimeter Institute, and so we packed the morning's mess back into a tidy parcel. We walked out into the cool sunshine.

"I believe that my obituary will say I was killed by a bicycle," he said idly as we crossed a street. Macular degeneration had left him able to see the moving outlines of large vehicles like cars and trucks, but bicycles were so slim, they eluded his perception.

Roger and I crossed paths at the institute later that afternoon. He pulled me aside and asked, "How are you planning to tell my story?" It was the first time he had ever inquired about the specifics of the manuscript.

Before I could answer, he carried on. "I want you to use these letters. I want people to know about Judith's role as my muse." After a lifetime of keeping their relationship private, he now wanted her credited.

In another conversation a few weeks later, he said, "The proof that there isn't a God is that if there were a God, there were many points in my life when he would have whispered in my ear, 'Wait for Judith.'"

In our subsequent interviews he rarely showed the same kind of vulnerability. He wasn't merely putting his guard up. Isolation affected him, made him less capable of feeling and less willing to speak about emotional matters.

In March 2020, when the world shut down to mitigate a global pandemic, Roger was already spending almost all his time alone. In lockdown, he became more solitary than ever. He spent the pandemic jotting down all the theories, ideas, and insights that had not yet made it from his private papers into peer-reviewed journals. His memories of his brother Jonathan and their games of Rock Paper Scissors pushed him toward a new theory of retroactive quantum collapse to explain how the human brain can move so quickly. He searched the cosmic microwave background for Hawking points whose existence would bolster his theory of a cyclic universe. He thought back to his letters to Judith as he revised and advanced twistor theory. He had a theorem for generalized dualization of conic sections that he had developed as an undergraduate student but had never got around to publishing. He was collaborating with a physicist at Boston University on a new book about Escher's mathematics. He tried to work on all of these and more at once.

"Everything is complicated, and everything is connected," he continued to say. But this mantra started to reflect more frustration than excitement.

His fingers and eyes and mind and elevated blood pressure seemed to be working against him. He was running short of time.

Shadows moved across his flat from morning to dusk. He could go days barely exchanging even a few words with another human being.

In solitude and silence, he spent more time exploring the far-distant future, when stars and galaxies had long since disappeared behind the event horizons of hyper-massive black holes; when entropy approached infinity and time and space lost their meaning; when Hawking radiation travelled endlessly through expanses of featureless, eventless sameness.

Roger drifted through the universe's heat death, filled not with a sense of despair or misery but with unshakeable belief in universal rebirth. He could still prove that as black holes burned the last fragments of matter and information into photonic energy, the geometry of space-time shifted and all of reality transformed from a cold, dead husk into a hot, dense pinpoint of unimaginable energy.

Roger spent day after day, night after night, sitting on the precipice between two universes, searching data for evidence that this transition zone really existed.

On October 6, 2020, Roger was in the shower when he heard his landline ringing. He answered it dripping wet and naked as an emperor.

The person on the other end asked him to hold for a call with a representative of the Nobel Committee. They took too long, though, and he hung up so he could dry and dress.

The representative rang back later in the day, finally able to deliver the message that Roger had won the 2020 Nobel Prize in Physics.

"It feels weird. It's very flattering and a huge honour and much appreciated," he told me a few hours after receiving the news. But, he added, "there is a certain negative aspect to it. People who get the Nobel Prize in some sense get *too* respected." With his work on conformal cyclic cosmology (CCC) still unfinished, he worried that he had won physics' top prize too *early* in his career.

His inbox filled with messages from journalists, colleagues, fans, old friends, and past lovers. The queen sent congratulations. Every branch of the BBC wanted their own interview. Douglas Hofstadter, author of *Gödel, Escher, Bach: An Eternal Golden Braid*, sent him a customized "ambigram," a type of calligraphic trickery he had invented. Right-side up, it read "Roger Penrose." Rotated 180°, the same characters spelled "Nobel Prize."

Salley Vickers rang him up a few days after the prize was announced. They hadn't spoken in more than three decades. All that time, she had been carrying around doubts and guilt about ending their relationship. "He had just won the Nobel Prize and was a little

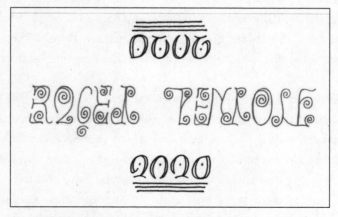

Douglas Hofstadter's congratulatory ambigram.
Courtesy of Douglas Hofstadter.

bit discombobulated. I said, 'I did regret it.' And he said, 'Oh, no, I think it was probably for the best. I went on and did all this work.' I used to think, 'Oh dear, how terrible. I behaved so badly to Roger.' But actually I think he was fine. I think I was the one who suffered," she said.[2]

She suggested they get together for lunch. He declined. He told me he was worried it might bother Vanessa if he visited with Salley. It seemed more likely, though, he just didn't want to reopen old connections and complications; Vanessa had nothing to do with it.

The Nobel Prize left Roger awash in a lifetime's worth of warmth and admiration. All he could think about was getting back to work.

He was working on a new paper about Hawking points and trying to reinvigorate the Penrose Institute. He envisioned a centre where theorists and experimentalists could work together, advancing and testing his theories of consciousness, cosmology, gravitizing quantum mechanics, twistor theory, and Newman-Penrose constants. He wanted the questions he had spent his life trying to answer to survive beyond his death.

Ivette Fuentes was one of a very small group of people with whom he maintained regular contact. "I can talk to him in a way I can't talk to anyone else. We get into this wave together where we just communicate so, so well. We go from criticising string theory to how close-minded people sometimes are, to how difficult it is to come up with something different or new," she said.

Roger had finally let go of his muse fantasy, though Ivette had to keep her guard up about how people viewed their relationship.

I once shared something on Facebook about Roger Penrose after we had given a talk together. A colleague from Mexico wrote a comment saying, "Oh well, I wish I was supported by him like that." It pisses me off because I worked so hard on my own to get where I am. In science sometimes, you have someone with a big name who has a group, and they support those people and make sure they get positions. They're like these science families if you want to see it like that. I never belonged to one. I was an orphan. Roger crossed paths with me, and this interaction and friendship and work relationship grew. But people say, "Oh she's just recognized because of Roger Penrose." Or even mean people implying other things. I was always clear with Roger I don't want that perception.

While conversations about the institute progressed, Roger worked furiously to solidify as many of his ideas as possible. "Covid has some good sides, being locked down. It gave me the chance to develop some new ideas I have had for a long time, the time to think what is wrong with quantum mechanics," he told the *London Times* in 2020.

He had filled his flat with all the things he loved most in the universe. Alongside his toys and games, hundreds of carefully curated books filled shelves in his office, bedroom, and living room—volumes on relativity and gravity, quantum mechanics, philosophy of science, impossible objects, consciousness, and the art of M. C. Escher. He kept his own works close by—the easiest volumes for him to reach from his office chair were the six-volume compilation of his first fifty years of peer-reviewed publications.

Tables, chairs, and other flat surfaces were strewn with new titles, jumbled among puzzles and toys. His cutting board and the cushion on his office chair both had Penrose tile motifs. A dilapidated leather suitcase held an eclectic collection of tactile brainteasers, including several of Lionel's handcrafted wooden puzzles. The games shelves held a Spirograph set, 2-D and 3-D jigsaw puzzles, a bag of Escher-inspired interlocking lizard tiles, and a handheld microscope. Every available space offered something fascinating to look at, play with, flip through, or puzzle over. He surrounded himself with the delicate, clever geometry that brought him happiness and peace of mind.

Living and working alone in his nineties was becoming more difficult.

Files disappeared from his overstuffed hard drive. Names disappeared from his memory. Making his coffee for the week one Friday, he spilled boiling water on his feet, inflicting second-degree burns. If he overexerted himself, his hip or back would ache for days.

Stephen Hawking died in March 2018. Wolfgang Rindler died in February 2019. Joan died later that year—Roger visited her in hospice to say goodbye. Ted Newman died in March 2021. Jonathan died the following November.

Vanessa periodically reminded him that there was a flat waiting for him closer to his family, where they could do a better job of looking after him. He didn't have to be so isolated. He insisted on staying where he was. He chose to be alone.

He battled with allies and opponents over how large Hawking points should be. He speculated anew about the possibility of the universe containing *naked singularities*—areas of infinite density not obscured behind an event horizon. Re-examining some of Escher's prints, he discovered new mathematics that seemed to predict new physical theories. He became fascinated with *octonions*—pairs of pairs of complex numbers that seemed better to describe the properties of forces and particles. These *hyper-complex numbers* renewed his hopes that twistor theory might still solve the shortcomings of quantum mechanics.

A few days before Ted died, Roger spoke with him via Zoom. They knew it would be their last conversation. "I made a vow to Ted to look at his latest work myself or get somebody else to do it, and to bring these N-P constants into an understanding of equations of motion," Roger said.[3]

Everything was complicated, everything was connected, and nothing was finished.

In 2022, we reached the end of a long day of in-person interviews. As I packed up my devices, he offhandedly observed how many scientists— not just him—seem to have "troubled marriages." He implied that a solitary life might be an inevitable consequence, a necessary price, for his kind of success. True, Wolfgang and Ted had had long, happy marriages. Then again, neither of them had won a Nobel Prize.

His tone was of justification rather than regret. He didn't see how it could be any other way. "There are things that I would like to have been different, but they're not necessarily plausible. I wish I'd been more at ease with women when I was young, and so therefore not got susceptible to Joan when she came into the picture."

I hadn't yet spoken to Eric or Toby at that time. Roger's sons were a pronounced non-presence in his life. He talked about his unease with women, but he also seemed put off by the thought of getting any closer to his sons. He had once told Salley his ideas were as dear to him as his children. In his nineties, when he considered his legacy, he made a clear choice of one over the other.

I think an important aspect of this is that the work is not finished. I think most people at my age, scientists, say, "OK, well, we've done good things or bad things, and this is it. You can go and retire or whatever it is." I don't have that feeling. I've got a very different feeling.

I keep trying to write this paper on split octonions, but there are so many distractions, and my eyesight isn't good, which means I'm extremely slow. If it was clear this is the right way to go and people were going to pick up on it and it would be in good hands, then I would say that's fine. I can go and enjoy myself and go to the movies or something instead. I don't feel that that's really happened.

I thought I knew all about twistor theory, and then this split octonian thing comes along and I realise there's something completely different, hiding there all the time. That needs thinking about. It's a completely new angle on twistor theory. And somehow, it's got to be tied up with CCC. Somehow, it's got to be tied in with general relativity.

I'm rambling on about this stuff just because it's part of the reason that I don't feel I can let it go. To me, it's very important. Of course I might just drop dead at any minute; then somebody else would have to take over. That could easily happen. But nevertheless, if I'm still around, to me it is the really important thing to do.[4]

When I first contacted Joan to interview her for this book, she had only recently heard that Roger and Vanessa had separated. Her first questions were, "How big is Roger's flat? Would he have room for his sons to come and stay with him for a visit?" She told me only later she had oesophageal cancer and didn't have long to live. She wanted their kids to have a better relationship with their father before she died.

I reached out to Christopher, Toby, and Eric many times, with no response. Eventually, I resigned myself to writing the book without their voices.

In July 2023, Roger sent me a Google Doc that Toby had shared with him, containing videos of the family at Blink Bonnie and other grainy clips digitized from old sixteen-millimetre films. I couldn't open the document. I wrote to Toby to request access. His response sparked an email conversation that led to interviews with both him and Eric.

Eric expressed anguish over Roger's treatment of Joan. "Looking back, I sensed a lasting impact it had on me," he said. "I hadn't actually thought about it much since that time, but Mum would say the problems in life I was having could be traced quite a lot back to my seeing his aggression against her—that I'd internalized the problem of witnessing it."[5]

Decades after the fact, Toby and Eric's childhood memories of Roger's physical aggression remained powerful. "I wouldn't feel right saying positive things if I didn't include that," Eric said. "It wouldn't feel right to me."

Both sons were still interested in finding a way to repair their relationship with their father.

"Despite the distance between me and Dad and a sense of rejection when I was growing up, I still felt there was some warmth and that he cared about me. More recently I feel we get on much better," Toby said.[6]

"Though I blame him for his aggression to mum, for me, I would say mostly he was good as a dad. Though I couldn't say there was generally a sense of a clear emotional connection," Eric said.[7]

Roger responded to his sons' memories with such evenness and neutrality as to appear indifferent: he only saw himself as the helpless fly caught in Joan's web.

I described to Roger his sons' anger and sense of abandonment and asked him if he had any "unfinished business"—things he'd like to say to them while he's still capable of doing so. "I feel my life is busy enough and if I get involved with them, it just distracts from other things," he said.

Speaking to Eric and Toby helped make sense of Roger's aversion to being involved in their lives. It wasn't just about the bother of dealing with other people's complicated emotions or the distraction from his work. If he let them in, he would have to face them, acknowledge his past actions, and take responsibility for his behaviour. If he kept them at a distance, he could go on believing he was the unfortunate victim of his own awkwardness and ill fate. The same would hold true for Christopher, Vanessa, Max, Salley, or anyone else who had paid a price to be a part of his world—or to be expelled from it.

Isolation provided him an escape from accountability.

At age ninety-two, he knew death could not be far away, be it from high blood pressure or a fatal bicycle collision. Roger appeared to be pre-emptively letting go, severing the last threads that tied him to family and friends.

Before he died, he chose to focus on making peace not with his sons or others he had shut out but with his unfinished ideas. Even his lifelong craving for a muse seemed to fade away.

I read a quote to him from a letter he had written Judith in 1976—after they had grown somewhat distant. "Women are still important to me in my work—even though I am much more independent of that now than before. And I am still sensitive to women's reactions to my work—even if it's quite superficial," he had written.

He laughed hearing these words again nearly half a century after he wrote them. "Well, I don't particularly recognize that in what I do now," he said.

A few weeks later I read him a quote from Douglas Scott, one of his critics on CCC: "It was only a prediction because Roger said it was a prediction." His anger percolated for several days after we spoke.

"I have to say that I'm very puzzled by Douglas Scott's reaction," he wrote to me. "When Eddington's expedition went to look for the displacement of the star field as predicted by general relativity, and the effect was actually seen, would Scott just have said, 'Oh well, that's just an effect. It doesn't have to have anything to do with Einstein's theory.' You could say that kind of thing about any observational test of a theory."

Physics debates animated and upset him in ways personal relationships did not. The anger and pain of his sons, his spouses, and his friends barely registered. He saw their sacrifices as unfortunate but necessary.

It's tempting to imagine that true genius—the kind of creative inspiration that fosters revolutionary ideas and transforms the world—requires this kind of sacrifice. Certain scientists, artists, and musicians are treated as so exceptional, so extraordinarily talented and driven, that the people around them must bend and suffer in the extreme so the genius can succeed.

It is easy to dismantle that type of seductive mythology. Roger would have been no less a physicist if he had been gentler with Joan or more respectful of Judith. Had he been more present with his sons or more receptive to Vanessa, had he made more room for the people who loved, understood, and supported him, his work might well have been more productive, more creative, and more successful.

Joan, Judith, Salley, and Vanessa, each in her own way, made Roger's career possible. They managed his everyday life, freeing him to journey into black holes and tinker with conformal geometry. Beyond offering the praise and encouragement he considered essential to his creative process, they were variously sounding boards and editors, career counsellors, confidantes, and muses, willing or not. Some sacrificed their own careers and put his needs ahead of their own. They offered intellectual, logistical, and emotional support. Joan, whom Lionel treated as a blight on the Penrose mystique, served a crucial role in developing Roger's belief in his own exceptional talents. He used their unhappy marriage as justification for fully embracing the work that established his career and won him the Nobel.

Recognizing that genius relies on connection, support, and relationships does not diminish Roger's undeniable talent or his extraordinary attunement to the unexpected simplicity of the universe. He gave the world much beauty and magic. He transformed the way we appreciate and understand reality. But his talents were not his alone. They grew from connection, shared joy and loss, sacrifice, anger, estrangement, and confusion—a context Roger often tried to resist and avoid.

He treated his own isolation and its impact on the people closest to him as inevitable and necessary, the price the universe required for him to succeed. But he was not a fly caught in a web or a bundle of particles drifting passively through space-time. His life and work resulted from his choices.

Like Penrose tiling, Roger's life continually accommodated one more piece, extending and varying the patterns and cycles. Unlike his tiles, human lives don't follow consistent rules. His constituent parts were in constant, unresolvable conflict with one another—brilliance and obliviousness, stubbornness and passivity, generosity and solipsism, joy and indifference.

He found inspiration in conversations and in the silence between conversations. His ideas seemed to come sometimes from nowhere and other times from painstaking effort.

The universe that produced him has its own gaps and inconsistencies: The irreconcilability of quantum mechanics and general relativity. The black hole information paradox. Numbers that are both real and imaginary. Music that is both happy and sad. Infinities of time, space, matter, and energy. Truths that can be known but not proven.

Roger treated every challenge—personal and professional—as a problem waiting to be solved. This drive connected him to ideas so vast and beautiful they swept him away. It also shut out friends and family, leaving them heartbroken over his indifference to their love.

Since childhood, he had harboured more questions than he could answer in a single lifetime. His every breakthrough led to new, more intriguing questions. Every particle in his body was still single-mindedly devoted to answering them, racing against the moment when those particles were no longer bound together into the conscious entity known as Roger Penrose.

There were too many pieces, too many fragments of thought that still needed to be placed in the proper context. In a finite lifetime—as in a finite book—there isn't room to put every last piece in place. The puzzle is never finished.

AUTHOR'S NOTE

Memories—even unreliable ones—enhance what can be learned through historical records. Roger told me the story of tricking his nanny with a semicircle of spinach dozens of times. I presented the details of this and other childhood recollections as he remembers them, even when it was impossible for me to fully fact-check them. The way he recalls the spinach incident helps explain how he understands himself and who he became; for that reason, such stories are crucial.

Roger cooperated extensively with this biography but did not have approval over any part of the manuscript or over whom I interviewed and what I researched. I undertook verification and fact-checking of his recollections wherever possible. I shared my analysis and impressions with him regularly. For five years, we spoke almost every week via videoconference or telephone. I also interviewed him in person many times, in England, Canada, and the United States. He was consistently willing to respond to my questions, even when the answers were difficult or painful. I am grateful for his generosity and openness and appreciate his willingness to respect my editorial independence.

There are points in this narrative where I ascribe motivations or inner monologue to Roger without quotes or footnotes. These judicious acts of extrapolation are based on our hundreds of hours of conversation, during which I came to know the nuances of his drives, convictions, and fascinations. We spent many hours flipping through his old notebooks, poring over his hand-drawn overheads, and sifting through boxes of old games and souvenirs. Many items brought back old memories or added details to stories I'd heard many times before.

Whenever possible, I ran Roger's accounts past other people to get their version of events. I didn't always have this option. I deeply regret that Wolfgang Rindler died before I had a chance to meet him. Judith Daniels and Margaret Penrose left behind a wealth of insight in their writing, but I wish I could have spoken with them. I interviewed Joan Penrose only once before she became too ill to talk. I particularly missed her presence after Toby and Eric Penrose described Roger's mistreatment of her. I made extensive effort to verify and clarify the sons' accounts and to understand as best I could what happened. I consider it a great loss, though, that Joan died before she had the opportunity to tell her own story.

Some people whose insights and stories I would have loved to include declined to participate or spoke on very limited terms. While I miss their voices, I respect their choices and have done my best to respect their privacy.

Among the dozens of people I did interview, I agreed to a small number of off-the-record conversations. Many of these served as "gut checks," helping me correct or confirm my own intuitions and analysis with people who knew Roger well during various chapters of his life—people who loved him and also understood his fallibilities.

Roger and others shared a wealth of unpublished letters, memoirs, and other personal documents, which provided intimate, unvarnished insight into Roger's life and work. Reading these materials alongside the six massive volumes of Roger's *Collected Works*, his many popular science books, and his abundant forewords, essays, and journal articles helped me deconstruct where his ideas came from and what he was thinking and feeling at key points in his career.

I visited many places where Roger had formative experiences. I explored "the Jungle" behind his childhood home in Colchester and stayed at Thorington Hall, tramping through the surrounding farmland and viewing spectacular starry skies. I stood on the cliffs in Bayfield, Ontario, where Roger watched outdoor movies, traced the route of the Victoria Street Road Clock in London, Ontario, and immersed myself in James Dalgety's obsessively beautiful puzzle museum (which had moved from Over Wallop to the village of Luppitt). In Pittsburgh, I spent several days with Roger and Ted Newman, as they reminisced about their long friendship and made plans for new research projects. At the Perimeter Institute in Waterloo, Ontario, and the Maths Institute at Oxford University, I watched admiring young physicists pepper him with questions and praise.

My first interview with Roger took place in October 2018. I submitted the final manuscript in April 2024. I didn't choose for this book to take that long, but I am fortunate to have had those years. I hope the results contribute to conversations about how science happens, where creative inspiration comes from, and who contributes and sacrifices to make works of genius possible.

ACKNOWLEDGEMENTS

I am deeply grateful to everyone who shared their stories and memories with me. Without their generosity and insight, this biography would not exist.

I am equally grateful to the many others who made this book possible, starting with my partner, Andrea Addario. Andrea read countless drafts of *The Impossible Man*, and I see her intelligence, integrity, and insight on every page. Her support for this book never wavered. I am lucky to share my life with her. Our sons, Abe and Isaac, whose curiosity, enthusiasm, and scepticism rival those of any scientist, inspired and motivated me through every stage of the book. My in-laws, Frank and Eleanor Addario, didn't live to see this book completed. I am thankful every day for the kindness and support they showed me when they were alive, as I am for the love and encouragement of my parents, Peter and Myra Barss.

For the first few years, this biography was an investment of time and energy with no guaranteed pay-off. In 2020, I was fortunate to be appointed science communicator in residence at York University in Toronto, a position that gave me time and resources to develop the book. I am grateful to York and to the creator of that programme, my friend Ray Jayawardhana.

In September 2021, I began a year-long fellowship at the Leon Levy Center for Biography at the City University of New York, funded by the Sloan Foundation. This fellowship not only provided me with the resources to focus exclusively on the book but also offered me a community of colleagues whose feedback and kindness continue to influence my writing and research. I am grateful to Kai Bird, director of the Leon Levy Center, for his editorial and logistical help. Thanks also to Thad Ziolkowski, the deputy director of the center, and to my stellar cofellows, Blair Brooks, Glenn Frankel, Helen Koh, Dan Nadel, Adam Schatz, and Rachel Swarns.

Thanks to Shaena Lambert, a good friend and a great novelist, who provided mentorship and feedback on the manuscript. Shaena has superb talent for setting scenes and bringing characters alive. I was lucky to be able to draw on her expertise.

Thanks to Michelle Frank and Elizabeth Greenwood, each of whom carried out crucial research for this book. A special thanks also to Ruth Gibson and others at the Bayview Historical Society in Bayview, Ontario. This group went far above and beyond to give me a sense of what this village and Blink Bonnie were like when Roger lived there. Similar appreciation to Rachel Strachan at the British National Trust, who helped me understand the history and architecture of Thorington Hall.

Journalist Katie McCormick fact-checked the science content. I am grateful for her expertise and attention to detail. Thanks also to my friend Shawn Micallef, a lecturer at the University of Toronto, for providing me invaluable access to the university's libraries.

While I was writing this book, several other people were working on projects with overlapping interests. Physicist Philip Stamp interviewed Roger extensively as part of a project to create an oral history of the past century of advances in cosmology. Science historian Dennis Lehmkuhl was involved in archiving some of Roger's papers. Michael Wright, who leads a charity called the Archive Trust for Research in Mathematical Sciences and Philosophy, was working to digitize and catalogue decades' worth of recordings of physics lectures, including many of Roger's old talks. I appreciated being able share ideas and resources with all of them.

Writing can be isolating, but through the Biographers International Organization I found a community of science biographers. Gabriella Kelly-Davies organized regular meetings where we discussed our work and commiserated. I deeply value these conversations.

I also appreciate the many friends, family members, and colleagues who were sounding boards, editors, and supporters at various stages of the book, including Safdar Abidi, Connie Addario, Susan Addario, Ingram Barss, Gerry Bergstein, Erin Bow, Emile Carrington, Leigh Cresswell, Naiara Perin Darim, Lucy Decoutere, Randy Dotinga, Joni Dufour, Aida Edemariam, Owen Egan, Dan Falk, David Farrar, Mike Fox, Corey Frost, Richard Gray, Geoffrey Hinton, Abby Levin, Farheen Mahmood, Andrew Mahon, Fay Meling Pao, Vince Pietropaolo, Sarah Polley, Siobhan Roberts, Rob Selkowitz, Ivan Semeniuk, Rosie Simmons, Seth Singer, Paul Szeptycki, Kathy Vey, Natasha Waxman, Tim Wayne, Karen Weingarten, and Nicki Weiss.

Finally, heartfelt thanks to my agent, Robin Straus, and my editors, T. J. Kelleher at Basic Books and James Nightingale at Atlantic Books. Their guidance and knowledge enriched the writing process and contributed greatly to the book's final form.

ART CREDITS

NOTES

CHAPTER 1: THE JUNGLE

1. The brutal language Lionel and contemporaneous researchers used to describe people with Down syndrome, autism, and other neurodivergent conditions reflected a strong societal predisposition toward eugenics. Lionel used this language but publicly opposed eugenic approaches. He devoted much of his career to using data and science to frame mental conditions as medical issues rather than moral or spiritual failings.

2. The survey was officially titled "A Clinical and Genetic Study of 1280 Cases of Mental Defect."

3. "Artistic and Kindly Quaker Who Made Oxhey Grange His Home," *News Shopper*, June 14, 2002.

4. Thomas Bewley, "Lionel Penrose, Fellow of the Royal Society," Cambridge.org.

5. She was born Sara Natanson but changed her name when she left Russia. Unpublished memoir of Margaret Penrose (hereafter referred to as "Margaret's memoir"), 3.

6. Tara Brookfield, *Our Voices Must Be Heard: Women and the Vote in Ontario* (Women's Suffrage and the Struggle for Democracy), University of British Columbia Press, 2018, p. 126.

7. Bedales: https://www.bedales.org.uk.

8. Roger Penrose, interview with author, January 19, 2022.

CHAPTER 2: UNEXPECTED SIMPLICITY

1. Joan Wedge, interview with author, November 10, 2020.

2. This version of the puzzle is reproduced from Alex Bellos, *Can You Solve My Problems*, The Experiment, 2017. The solution rests on the premise that no statement is superfluous. Using a series of if-then propositions, Bellos demonstrates that only one order of choosing—Low, Critic, Y. Y.—ensures all three statements provide relevant information. Not only is the original challenge

beyond the abilities of most recreational puzzle enthusiasts, but the explanation of the solution also leaves most readers in the dark. Newman's puzzles were designed for an extremely select group, which included Lionel Penrose in the 1930s and Roger Penrose some years later.

3. David Ross, ed., "Thorington Hall," Britain Express, https://www.britainexpress.com/attractions.htm?attraction=3704.

4. Ocean Travelers, *New York Times* notice, April 14, 1939.

5. Roderick Barman, ed., *Safe Haven: The Wartime Letters of Ben Barman and Margaret Penrose, 1940–1943*, McGill Queens University Press, 2018, p. 34.

6. Roderick Barman, ed., *Safe Haven: The Wartime Letters of Ben Barman and Margaret Penrose, 1940–1943*, McGill Queens University Press, 2018, p. 3.

7. Margaret's memoir, p. 25.

8. Roger Penrose, *The Role of Aesthetics in Pure and Applied Mathematical Research* (Bulletin of the Institute of Mathematics and Its Applications #10), Institute of Mathematics and Its Applications, 1974, pp. 266–271.

9. The Role of Aesthetics in Pure and Applied Mathematical Research IMA 1974.pdf (volume 2 of collected works), p. 742.

10. Joan Penrose used this word to describe the relationship between the boys.

11. Roger Penrose, interview with author, August 14, 2021.

12. "Blink Bonnie," ChestofBooks.com. The region of Ontario that contains London and Bayfield also has a hamlet of the same name, but the cottage Blink Bonnie appears to have been independently named.

13. *Perils of Nyoka*, Republic Films, 1942; see *"Perils of Nyoka,"* Wikipedia, https://en.wikipedia.org/wiki/Perils_of_Nyoka.

14. Roger Penrose, interview with author, January 30, 2021.

15. J. B. S. Haldane to Lionel Penrose, August 18, 1943.

16. Roger Penrose, interview with author, April 10, 2021.

CHAPTER 3: THE ARROW OF TIME

1. Roger Penrose, interview with author, December 11, 2019.

2. Roger Penrose, interview with author, January 30, 2021.

3. Roger Penrose, interview with author, January 14, 2021.

4. Roger Penrose, interview with author, November 24, 2018.

5. Roger Penrose, interview with author, November 24, 2018.

6. Roger Penrose, interview with author, November 24, 2018.

7. Roger Penrose, letter to Judith Daniels, October 21, 1972.

8. Roger Penrose, letter to Judith Daniels, October 21, 1972.

9. Roger Penrose, interview with author, August 28, 2022.

10. Roger Penrose, interview with author, February 18, 2023.

11. Roger Penrose, interview with author, April 10, 2021.
12. Roger Penrose, interview with author, April 10, 2021.
13. Roger Penrose, interview with author, October 30, 2021.
14. Roger Penrose, interview with author, September 16, 2023.

CHAPTER 4: THE IMPOSSIBLE TRIANGLE

1. "The Congress was attended by 1553 full members and 567 associate members": "1954 ICM—Amsterdam," MacTutor, https://mathshistory.st-andrews.ac.uk.

2. *Proceedings of the International Congress of Mathematicians*, Vol. 1 (1954): p. 141. Translation available at "1954 ICM—Amsterdam," MacTutor, https://mathshistory.st-andrews.ac.uk/ICM/ICM_Amsterdam_1954/#sect2.

3. *Proceedings of the International Congress of Mathematicians*, Vol. 1 (1954): p. 143.

4. *Proceedings of the International Congress of Mathematicians*, Vol. 1 (1954): pp. 144–145.

5. *Proceedings of the International Congress of Mathematicians*, Vol. 1 (1954): pp. 144–145, p. 162.

6. Roger Penrose, interview with author, October 23, 2021.

7. *Proceedings of the International Congress of Mathematicians*, Vol. 1 (1954): p. 158. Roger misremembered the show being at the Van Gogh Museum.

8. Roger Penrose, introduction to paper "Impossible Objects: A Special Type of Visual Illusion," in *Roger Penrose: Collected Works*, Vol. 1: *1953–1967*, Oxford University Press, 2011, p. 355.

9. "Roger Penrose," interview by Alan Lightman, American Institute of Physics Oral History, January 24, 1989.

10. Roger Penrose, interview with author, September 17, 2022.
11. Roger Penrose, interview with author, September 17, 2022.
12. Roger Penrose, interview with author, September 17, 2022.
13. Roger Penrose, interview with author, November 20, 2021.
14. J. A. Todd, "Report on the Dissertation by R Penrose: Tensor Methods in Algebraic Geometry," Cambridge Archives, February 1957.
15. Roger Penrose, interview with author, November 10, 2020.
16. Roger Penrose, interview with author, November 10, 2020.
17. Roger Penrose, interview with author, November 10, 2020.
18. Joan Penrose, interview with author, February 6, 2019.
19. Joan Penrose, interview with author, February 6, 2019.
20. Joan Penrose, interview with author, February 6, 2019.
21. Roger Penrose, interview with author, November 10, 2020.
22. Roger Penrose, interview with author, November 24, 2020.

23. Roger Penrose, interview with author, November 24, 2018.

24. Joan Penrose, interview with author, February 6, 2019.

25. Roger Penrose, interview with author, February 2, 2019.

26. Roger Penrose, interview with author, February 2, 2019.

27. Roger Penrose, interview with author, November 14, 2020.

28. Roger Penrose, interview with author, November 14, 2020.

29. L. S. Penrose and R. Penrose, "Impossible Objects: A Special Type of Visual Illusion," *Journal of Psychology* 49 (1958): pp. 31–33.

30. Lionel Penrose, letter to Escher, December 7, 1961. Courtesy of the Escher Foundation.

31. L. S. Penrose and Roger Penrose, "Puzzles for Christmas," *New Scientist* (December 1958): pp. 1580–1588.

32. Roger Penrose, interview with author, August 20, 2022.

33. Roger Penrose, "General Relativity in Spinor Form," *Les Theories relativistes de la gravitation*, ed. A. Lichnerowicz and M. A. Tonnelat, Centre National de la Recherche Scientifique, 1982, pp. 429–432.

34. Roger Penrose, "The Apparent Shape of a Relativistically Moving Sphere," *Proceedings of the Cambridge Philosophical Society* 55, no. 1 (1959): pp. 137–139.

CHAPTER 5: BLINK BONNIE

1. The house was at 15361 Indiana Avenue, according to a letter from Joan to Lionel and Margaret, dated June 28, 1960, in the University College London's L. S. Penrose Papers archive (hereafter cited as Penrose Papers), which was built in segments through the late 1950s and early 1960s.

2. Roger Penrose, letter to Lionel Penrose, June 28, 1960, Penrose Papers.

3. Details from two letters: Roger to Lionel, June 28, 1960, Penrose Papers; Joan to Lionel, October 18, 1960, Penrose Papers.

4. Roger Penrose, letter to Lionel Penrose, June 28, 1960, Penrose Papers.

5. "Bayfield Cottages and Homes and the People Who Owned and Lived in Them," p. 45, courtesy of the Bayfield Historical Society.

6. Joan Penrose to Lionel Penrose, October 18, 1960, L. S. Penrose Papers Wellcome Collection, https://wellcomecollection.org/works/quedyxap.

7. Joan Penrose, letter to Lionel Penrose, October 18, 1960, Penrose Papers, p. 98 in Correspondence P.

8. Charles W. Misner and Kip S. Thorne, "Preface," in *Gravitation*, Princeton University Press, 2017.

9. Roger Penrose, interview with author, February 8, 2019.

10. Roger Penrose, *Collected Works*, Vol. 1: *1953–1967*, Oxford University Press, 2011, p. 369.

11. Joan Penrose, interview with author, February 6, 2019.

12. Roger Penrose, interview with author, July 24, 2021.

13. Roger Penrose, interview with author, November 10, 2020.

14. Roger Penrose, interview with author, January 14, 2023.

15. Ted Newman, interview with author, December 11, 2018.

16. Roger Penrose, interview with author, December 11, 2019.

17. Roger Penrose, interview with author, December 11, 2019.

CHAPTER 6: JABLONNA AND BAARN

1. Updated introduction to "An Approach to Gravitational Radiation by a Method of Spin Coefficients," in *Roger Penrose: Collected Works*, Vol. 1: *1953–1967*, Oxford University Press, 2011, p. 421.

2. Rindler was part of the UK-organized rescue of Jewish children known as the *Kindertransport*. About 10,000 children, mostly Jewish, came to England without their families in the months leading up to World War II.

3. Wolfgang Rindler, interview with Jacob Moldenhauer (unpublished), May 17, 2017, and October 26, 2016.

4. Joan Wedge, interview with author, November 10, 2020.

5. Andrzej Trautman and Donald Salisbury, "Memories of My Early Career in Relativity Physics," interview recorded June 24 and 28, 2016, arXiv, https://arxiv.org/abs/1909.12165.

6. Ted Newman, interview with author, November 28, 2018.

7. Roger Penrose, interview with author, December 20, 2022.

8. Wolfgang Rindler, "After Dinner Speech," tribute to Ivor Robinson, delivered at University of Texas at Dallas, May 7, 2017, available at http://artemis.austincollege.edu/acad/physics/dsalis/RobinsonMemorial/Rindler%20After%20Dinner%20Speech.pdf.

9. Roger Penrose, unrecorded interview with author, December 1, 2022.

10. Lawrence Krauss, interview with Roger Penrose, March 24, 2022, available at "The Origins Podcast: Roger Penrose," *Critical Mass* (Substack), March 25, 2022, https://lawrencekrauss.substack.com/p/ad-free-the-origins-podcast-roger?s=w.

11. From a letter from M. C. Escher to C. V. S. Roosevelt, March 8, 1962. Quoted in J. Taylor Hollist, "Escher Correspondence in the Roosevelt Collection," *Leonardo* 24, no. 3 (1991): pp. 329–331.

CHAPTER 7: ASSASSINATION

1. Joan Penrose, interview with author, October 11, 2020.

2. Roger Penrose, multiple interviews with author.

3. Interviews with Roger Penrose, plus Englebert Shucking, "The First Texas Symposium on Relativistic Astrophysics," *Physics Today* 42, no. 8 (1989).

4.   Engelbert Shucking, "The First Texas Symposium on Relativistic Astrophysics," *Physics Today* 42, no. 8 (1989): p. 46.

5.   Roger Penrose, unrecorded interview with author, December 1, 2022.

6.   Roger Penrose, interview with author, October 9, 2021.

7.   Roger Penrose, interview with author, June 19, 2019.

8.   In 1966, an architectural engineering student barricaded himself on the observation deck of the Tower with a rifle and killed fourteen people, including one member of the physics department. Roger had already moved on at this point.

9.   "The Big Enormous Building," *The UT History Corner*, February 10, 2017.

10.   Roger Penrose, letter to Judith Daniels, October 10, 1972.

11.   Roger Penrose, letter to Judith Daniels, October 7, 8, 10, 1972.

12.   "A Conversation with Wolfgang Rindler from Aug. 13, 2009," video posted to YouTube by Bass School at UT Dallas, September 1, 2021.

13.   "KRLD-TV Raw Footage from the Dallas Trade Mart on November 22, 1963," video posted to YouTube by David Von Pein's JFK Channel, October 28, 2014.

14.   Brian Clegg, "20 Amazing Facts About the Human Body," *The Guardian*, January 27, 2013.

15.   "A Conversation with Wolfgang Rindler from Aug. 13, 2009," video posted to YouTube by Bass School at UT Dallas, September 1, 2021.

16.   "Unease, Then Shock Followed Wait at Trade Mart for JFK Speech That Never Came," *Dallas Morning News*, November 1, 2013.

17.   Zsuzsanna Ozsváth, archival interview, courtesy of the William E. Cooper Collection/The Sixth Floor Museum at Dealey Plaza.

18.   "A Conversation with Wolfgang Rindler from Aug. 13, 2009," video posted to YouTube by Bass School at UT Dallas, September 1, 2021.

19.   Roger Penrose, interview with author, June 19, 2019.

20.   Roger Penrose, interview with author, January 19, 2022.

21.   Zsuzsanna Ozsváth, interview with author, November 1, 2021.

22.   Roger Penrose, interview with author, June 19, 2019.

23.   Roger Penrose, *On the Origins of Twistor Theory. Gravitation and Geometry: A Volume in Honour of Ivor Robinson* (Monographs and Textbooks in Physical Science Lecture Notes 4), Bibliopolis Edizioni di Filosofia e Scienze, Naples, 1987, pp. 341–361.

24.   Michael Atiyah, Maciej Dunajski, and Lionel J. Mason, "Twistor Theory at Fifty: From Contour Integrals to Twistor Strings," *Proceedings of the Royal Society of London A* 473 (2017): 20170530.

25.   Roger Penrose, interview with author, January 19, 2022.

26.   Roger Penrose, unrecorded interview with author, December 1, 2022.

CHAPTER 8: THE SKY IN A DIAMOND

1. Roger Penrose, interview with author, February 2. 2019.

CHAPTER 9: THE PENROSE SINGULARITY THEOREM

1. Roger Penrose, interview with author, October 6, 2020.
2. Kip Thorne, interview with author, April 8, 2022.
3. Wolfgang Rindler, "After Dinner Speech," tribute to Ivor Robinson, delivered at University of Texas, Dallas, May 7, 2017.
4. Roger Penrose, interview with author, July 13, 2019.
5. Roger Penrose, interview with author, April 10, 2021.
6. Roger Penrose, interview with author, January 27, 2019.
7. Kip Thorne, interview with author, April 8, 2022.

CHAPTER 10: HAWKING AND PENROSE

1. Roger Penrose, interview with author, January 13, 2020.
2. Roger Penrose, interview with author, January 19, 2022.
3. Roger Penrose, interview with author, March 24, 2022.
4. Natalie Wolchover, "How One Man Waged War Against Gravity," *Popular Science*, March 15, 2011.
5. The Brans-Dicke theory is an alternate description of gravity, a competitor of general relativity in which gravity is not a constant but a variable that changes over time and in different places. Stephen Hawking and Roger Penrose, "On Gravitational Collapse and Cosmology," Gravity Research Foundation, 1968.
6. Stephen Hawking and Roger Penrose, "On Gravitational Collapse and Cosmology," Gravity Research Foundation, 1968.
7. Roger Penrose, "Introduction to *The Singularities of Gravitational Collapse and Cosmology*," in *Roger Penrose: Collected Works*, Vol. 2: *1968–1975*, Oxford University Press, 2011, p. 267.
8. Paul Tod, interview with author, February 1, 2019.
9. Roger Penrose, interview with author, March 24, 2022.
10. Roger Penrose, interview with author, March 24, 2022.
11. Roger Penrose, interview with author, March 12, 2022.

CHAPTER 11: JUDITH DANIELS

1. Michel Baranger and Raymond A. Sorensen, "The Size and Shape of Atomic Nuclei," *Scientific American*, August 1969, pp. 58–73.
2. John N. Babcall, "Neutrinos from the Sun," *Scientific American*, September 1969, pp. 29–37.

3. Leo Goldberg, "Ultraviolet Astronomy," *Scientific American*, June 1969, pp. 92–102.

4. H. W. Hayden, R. C. Gibson, and J. H. Brophy, "Superplastic Metals," *Scientific American*, March 1969, pp. 28–35.

5. Von R. Eshleman, "The Atmospheres of Mars and Venus," *Scientific American*, March 1969, pp. 78–91.

6. Roger Penrose, letter to Judith Daniels, July 31, 1971.

7. Roger Penrose, letter to Judith Daniels, September 29, 1972.

8. Roger Penrose, letter to Judith Daniels, September 13, 1971.

9. Roger Penrose, letter to Judith Daniels, September 13, 1971.

10. Roger Penrose, letter to Judith Daniels, September 13, 1971.

11. Judith Daniels, letter to Roger Penrose, September 3, 1971.

12. Helen Hulson, unrecorded interview with author, May 22, 2019.

13. Roger Penrose, letter to Judith Daniels, July 31, 1971.

14. Roger Penrose, letter to Judith Daniels, August 17, 1971.

15. Roger Penrose, letter to Judith Daniels, August 17, 1971 (mailed in same envelope as two other letters).

16. Roger Penrose, letter to Judith Daniels, August 25, 1971.

17. Roger Penrose, letter to Judith Daniels, August 25, 1971.

18. Judith Daniels, letter to Roger Penrose, September 3, 1971.

19. Roger Penrose, letter to Judith Daniels, September 7, 1971.

20. Roger Penrose, letter to Judith Daniels, September 7, 1971.

21. Roger Penrose, letter to Judith Daniels, September 7, 1971.

22. Judith Daniels, letter to Roger Penrose, September 8, 1971.

23. Judith Daniels, letter to Roger Penrose, September 8, 1971.

24. Judith Daniels, letter to Roger Penrose, September 8, 1971.

25. Roger Penrose and Wolfgang Rindler, *Spinors and Space-Time*, Cambridge University Press, 1984, p. 424.

26. Roger Penrose and Wolfgang Rindler, *Spinors and Space-Time*, Cambridge University Press, 1984, p. 425.

27. Judith Daniels, letter to Roger Penrose, September 8, 2021.

28. Judith Daniels, letter to Roger Penrose, September 8, 2021.

29. Roger Penrose, letter to Judith Daniels, September 13, 1971.

30. Roger Penrose, letter to Judith Daniels, September 13, 1971.

31. Judith Daniels, letter to Roger Penrose, September 26, 1971.

32. Roger Penrose, letter to Judith Daniels, September 29, 1971.

CHAPTER 12: TRIESTE

1. Ray Sachs, email to author, August 19, 2020.

2. Roger Penrose, interview with author, October 9, 2021.

3. John Wheeler, letter to D. S. Jones, January 3, 1972, Roger Penrose Correspondence 1959–1973, John Archibald Wheeler Papers, American Philosophical Society; Emma Brown, "Ann E. Ewing, Journalist First Reported Black Holes," Boston.com, August 3, 2010.

4. John Wheeler to D. S. Jones, January 3, 1972, Roger Penrose Correspondence 1959–1973, John Archibald Wheeler Papers, American Philosophical Society.

5. John Wheeler to D. S. Jones, January 3, 1972, Roger Penrose Correspondence 1959–1973, John Archibald Wheeler Papers, American Philosophical Society.

6. John Wheeler to Roger Penrose, Roger Penrose Correspondence 1959–1973, John Archibald Wheeler Papers, American Philosophical Society.

7. Emma Brown, "Ann E. Ewing, Journalist First Reported Black Holes," Boston.com, August 3, 2010.

8. Roger Penrose, "Black Holes," *Scientific American* 226, no. 5 (May 1972), pp. 38–56.

9. Roger Penrose, "Black Holes," *Scientific American* 226, no. 5 (May 1972), pp. 38–56.

10. Roger Penrose, interview with author, October 2, 2021.

11. Roger Penrose, interview with author, January 12, 2019.

12. Roger Penrose, letter to Judith Daniels, August 22, 1972.

13. Toby Penrose, interview with author, December 14, 2023, and email to author, January 23, 2024.

14. Toby Penrose, interview with author.

15. Eric Penrose, interview with author, September 5, 2023.

16. Joan died before Toby and Eric shared their stories. I regret not having had the opportunity to speak to her directly about these events.

17. Roger Penrose, interview with author, September 25, 2023.

18. Toby Penrose, email to author, November 22, 2023.

19. Toby Penrose, interview with author, September 20, 2023.

20. R. Penrose and M. A. H. MacCallum, "Twistor Theory: An Approach to the Quantisation of Fields and Space-Time," *Physics Reports* 6, no. 4 (February 1, 1973): pp. 241–315.

21. Roger Penrose, letter to Judith Daniels, August 22, 1972.

22. Roger Penrose, letter to Judith Daniels, August 22, 1972.

23. Roger Penrose, letter to Judith Daniels, August 6, 1972.

24. Roger Penrose, letter to Judith Daniels, August 6, 1972.

25. Roger Penrose, letter to Judith Daniels, August 17, 1972.

26. Roger Penrose, letter to Judith Daniels, August 7, 1972.

27. Roger Penrose, letter to Judith Daniels, August 22, 1972.

28. Roger Penrose, letter to Judith Daniels, August 22, 1972.

29. Roger Penrose, letter to Judith Daniels, August 22, 1972.

30. Roger Penrose, letter to Judith Daniels, August 22, 1972.

31. Roger Penrose, letter to Judith Daniels, August 22, 1972.

32. Roger Penrose, interview with author, May 2, 2020 ("I would write letters to Judith in the middle of the night, and then walk off to the post box in my bare feet to mail the letter at three in the morning").

33. P. Barut, *Group Theory in Non-linear Problems: Lectures Presented at the NATO Advanced Study Institute on Mathematical Physics, Held in Istanbul, Turkey, August 7–18, 1972*, Springer Science & Business Media, 2012.

34. Roger Penrose, letter to Judith Daniels, August 27, 1972.

35. Roger Penrose, letter to Judith Daniels, August 27, 1972.

36. Roger Penrose, letter to Judith Daniels, October 19, 1972.

37. Roger Penrose, letter to Judith Daniels, October 1, 1972.

38. Roger Penrose, letter to Judith Daniels, August 30, 1972.

39. Roger Penrose, letter to Judith Daniels, August 27, 1972.

40. Roger Penrose, letter to Judith Daniels, October 18–19, 1972. (There is reason to believe this letter may actually have been written on September 18–19, 1972.)

41. Roger Penrose, letter to Judith Daniels, October 18–19, 1972. (There is reason to believe this letter may actually have been written on September 18–19, 1972.)

42. Roger Penrose, letter to Judith Daniels, September 6, 1972.

43. Roger Penrose, letter to Judith Daniels, September 6, 1972.

44. Roger Penrose, letter to Judith Daniels, September 6, 1972.

45. Roger Penrose, letter to Judith Daniels, October 19, 1972.

46. Judith Daniels, letter to Roger Penrose, September 12, 1972.

47. Judith Daniels, letter to Roger Penrose, September 12, 1972.

48. Judith Daniels, letter to Roger Penrose, September 12, 1972.

49. Judith Daniels, letter to Roger Penrose, September 12, 1972.

50. Judith Daniels, letter to Roger Penrose, September 20, 1972.

51. Judith Daniels, letter to Roger Penrose, October 27, 1972.

52. Roger Penrose, letter to Judith Daniels October 8, 1972.

53. Roger Penrose, letter to Judith Daniels, October 7, 1972; Roger Penrose, letter to Judith Daniels, October 8, 1972; Roger Penrose, letter to Judith Daniels, October 9, 1972.

54. Roger Penrose, letter to Judith Daniels, October 7, 1972; Roger Penrose, letter to Judith Daniels, October 8, 1972; Roger Penrose, letter to Judith Daniels, October 9, 1972.

55. Roger Penrose, interview with author, October 16, 2021.

56. Roger Penrose, letter to Judith Daniels, December 17, 1972.

57. Roger Penrose, letter to Judith Daniels, December 17, 1972.

58. Roger Penrose, interview with author, October 16, 2021.

CHAPTER 13: APERIODIC

1. Toby Penrose, interview with author, August 2, 2023.
2. Roger Penrose, interview with author, December 18, 2021.
3. Roger Penrose, interview with author, December 18, 2021.
4. Roger Penrose, interview with author, November 19, 2021.
5. Roger Penrose, letter to Judith Daniels, March 12, 1974.
6. Roger Penrose, interview with author, November 20, 2021.
7. Roger Penrose, Interview with author, November 19, 2021.
8. Judith Daniels, letter to Roger Penrose, June 17, 1973.
9. Roger Penrose, letter to Judith Daniels, June 24, 1973 (written from Vienna where he was working with Rindler on *Spinors and Space-Time*).
10. Roger Penrose, letter to Judith Daniels, June 24, 1973.
11. Judith Daniels, letter to Roger Penrose, July 2, 1973.
12. Judith Daniels, letter to Roger Penrose, July 6, 1973.
13. Judith Daniels, letter to Roger Penrose, July 6, 1973.
14. Roger Penrose, letter to Judith Daniels, February 1, 1973.
15. Robert Berger, "The Undecidability of the Domino Problem," *Memoirs of the American Mathematical Society*, No. 66. Advanced Mathematical Society, 1966.
16. Judith Daniels, letter to Roger Penrose, February 26, 1974.
17. Roger Penrose, letter to Judith Daniels, March 5, 1974.
18. Roger Penrose, letter to Judith Daniels, March 5, 1974.
19. Roger Penrose, letter to Judith Daniels, March 12, 1974.
20. Roger Penrose, letter to Judith Daniels, March 18, 1974.
21. Judith Daniels, letter to Roger Penrose, March 22, 1974.
22. Judith Daniels, letter to Roger Penrose, March 22, 1974.
23. Roger Penrose, letter to Judith Daniels, March 25, 1974.
24. Simon Kochen, interview with author, November 19, 2021.
25. Roger Penrose, letter to Judith Daniels. April 12, 1974.

CHAPTER 14: FAITH

1. Martin Gardner, "Extraordinary Nonperiodic Tiling That Enriches the Theory of Tiles," *Scientific American*, January 1977, pp. 110–121.
2. Judith Daniels, letter to Roger Penrose, December 10, 1974.
3. Roger Penrose, letter to Judith Daniels, May 3, 1975.
4. Siobhan Roberts, *Genius at Play: The Curious Mind of John Horton Conway*, Bloomsbury, 2015, p. 232.
5. Roger Penrose, letter to Judith Daniels, January 6, 1976.
6. Identity withheld, letter to Roger Penrose, January 3, 1976.
7. Judith Daniels, letter to Roger Penrose, March 13, 1976.

8. Roger Penrose, letter to Judith Daniels, March 22, 1976.

9. Jed Macosko, "Top Influential Physicists Today," Academic Influence, updated July 28, 2023.

10. Lee Smolin, interview with author, October 14, 2019.

11. Roger Penrose, letter to Judith Daniels, May 15, 1976.

12. Roger Penrose, letter to Judith Daniels, May 15, 1976.

13. Roger Penrose, letter to Judith Daniels, September 10, 1976.

14. Roger Penrose, letter to Judith Daniels, September 10, 1976.

15. Judith Daniels, letter to Roger Penrose, October 13, 1976.

CHAPTER 15: THE GAUGUIN DECISION

1. Salley Vickers, interview with author, March 10, 2023.

2. Salley Vickers, letter to Roger Penrose, May 20, 1978.

3. Roger Penrose, interview with author, March 25, 2023.

4. Roger Penrose, letter to Salley Vickers, March 7, 1985.

5. Salley Vickers, interview with author, March 10, 2023.

6. Salley Vickers, interview with author, March 10, 2023.

7. Salley Vickers, interview with author, March 10, 2023.

8. "Gauguin's Children," Gauguin Gallery, gauguingallery.com.

9. Roger Penrose, letter to Salley Vickers, undated.

10. Salley Vickers, email to author, April 5, 2023.

CHAPTER 16: UNDERSTANDING

1. David Deutsch, interview with author, February 14, 2022.

2. Tristan Needham, interview with author, February 10, 2022.

3. Tristan Needham, interview with author, February 10, 2022.

4. Tristan Needham, interview with author, February 10, 2022.

5. Roger Penrose, interview with author, July 23, 2022.

6. Tristan Needham, interview with author, February 10, 2022.

7. David Deutsch, interview with author, February 14, 2022.

8. Tristan Needham, interview with author, February 10, 2022.

9. Vanessa Penrose, email to author, September 19, 2023.

10. Tristan Needham, interview with author, February 10, 2022.

11. Roger Penrose, interview with author, January 20, 2019.

12. John McCarthy, "Review of *The Emperor's New Mind*," Formal Reasoning Group, April 7, 1998.

13. Daniel Dennett, "Review of Penrose, *The Emperor's New Mind*," *Times Literary Supplement*, September 19, 1989.

14. Timothy Ferris, "How the Brain Works, Maybe," *New York Times*, November 19, 1989.

15. Bryan Keating, Stuart Hameroff, and Roger Penrose, "Sir Roger Penrose & Stuart Hameroff: What Is Consciousness? Part 1," video posted to YouTube by Dr Brian Keating, August 5, 2022.

16. Roger Penrose, interview with author, September 17, 2022.

17. Daniel C. Dennett, interview with author, September 13, 2022.

18. Daniel C. Dennett, interview with author, September 13, 2022.

19. Steve Volk, "Can Quantum Physics Explain Consciousness? One Scientist Thinks It Might," *Discover*, March 1, 2018.

20. Roger Penrose, interview with author, September 17, 2022.

21. Steve Volk, "Can Quantum Physics Explain Consciousness? One Scientist Thinks It Might," *Discover*, March 1, 2018.

22. Lionel Mason, interview with author, February 1, 2019.

23. Vanessa Penrose, email to author, September 23, 2022.

24. Lionel Mason, interview with author, February 1, 2019.

25. Lionel Mason, interview with author, February 1, 2019.

26. Roger Penrose, *The Emperor's New Mind*, Oxford University Press, 1994, p. 8.

27. Roger Penrose, *The Emperor's New Mind*, Oxford University Press, 1994, p. 1.

28. Roger Penrose, *The Emperor's New Mind*, Oxford University Press, 1994, p. 3.

29. Roger Penrose, *The Emperor's New Mind*, Oxford University Press, 1994, p. 47.

30. Roger Penrose, *The Emperor's New Mind*, Oxford University Press, 1994, pp. 365, 367.

31. Hilary Putnam, "The Best of All Possible Brains?," *New York Times*, November 20, 1994.

32. Roger Penrose, Abner Shiony, Nancy Cartwright, and Stephen Hawking, *The Large, the Small, and the Human Mind*, Cambridge University Press, 1997.

33. Roger Penrose, interview with author, September 17, 2022.

34. Ivette Fuentes, interview with author, June 27, 2022.

35. Vanessa Penrose, email to author, March 30, 2022.

36. Roger Penrose, interview with author, January 19, 2022.

37. "Kleenex Art That Ended in Tears," *The Independent*, April 11, 1997.

38. Vanessa Penrose, email to author, March 23, 2022.

39. Roger Penrose, interview with author, September 17, 2022.

CHAPTER 17: CYCLES

1. Roger Penrose, interview with author, October 28, 2023.

2. Roger Penrose, "Singularities and Time-Asymmetry," in *General Relativity:*

An Einstein Century Survey, ed. S. W. Hawking and W. Israel, Cambridge University Press, 1979, pp. 581–638.

3. Roger Penrose, interview with author, October 29, 2022.

4. Roger Penrose, preface to *Cycles of Time*, Knopf Doubleday, 2011.

5. Roger Penrose, interview with author, October 29, 2022.

6. Roger Penrose, *Cycles of Time: An Extraordinary New View of the Universe*, The Bodley Head, 2010, chap. 3.4.

7. Peter Woit, "Cycles of Time," *Not Even Wrong*, May 27, 2011.

8. Roger Penrose, interview with author, February 8, 2019.

9. Roger Penrose, interview with author, January 24, 2019.

10. Douglas Scott, interview with author, April 15, 2023.

11. Douglas Spergel, interview with author, September 22, 2022.

12. Aja Romano, "How Do You Solve a Problem like Joe Rogan," Vox, February 23, 2022; Jonathan Jarry, "Science vs. Joe Rogan," McGill, November 20, 2021.

13. Lionel Mason, interview with author, February 1, 2019.

14. Daniel Shechtman, interview with author for "Impossible Cookware and Other Triumphs of the Penrose Tile," *Nautilus*, April 25, 2014.

CHAPTER 18: FANTASY

1. Roger Penrose, interview with author, October 29, 2022.

2. Graham Farmello, "Fashion, Faith and Fantasy in the New Physics of the Universe by Roger Penrose—Review," *The Guardian*, November 12, 2016.

3. Roger Penrose, *Fashion, Faith, and Fantasy in the New Physics of the Universe*, Princeton University Press, 2016.

4. Ivette Fuentes, interview with author, June 27, 2022.

5. Ivette Fuentes, interview with author, June 27, 2022.

6. "Consciousness Central 2017—Day 4 with guests Sir Roger Penrose and James Tagg and Elaine Chew," video posted to YouTube by Conscious Pictures, June 15, 2017.

7. Lionel Mason, interview with author, February 2, 2019.

8. Roger Penrose, interview with author, October 1, 2022.

9. Jeffrey Mervis, "What Kind of Researcher Did Sex Offender Jeffrey Epstein Like to Fund?," *Science*, September 19, 2019.

10. Ivette Fuentes, interview with author, June 27, 2022.

11. Ivette Fuentes, interview with author, June 28, 2022.

12. Ivette Fuentes, interview with author, June 28, 2022.

CHAPTER 19: MORE TIME TO THINK

1. Roger Penrose, interview with author, February 8, 2019.

2. Salley Vickers, interview with author, March 10, 2023.

3. Roger Penrose, interview with author, September 17, 2022.
4. Roger Penrose, interview with author, March 24, 2022.
5. Eric Penrose, interview with author, September 5, 2023.
6. Toby and Eric Penrose, joint interview with author, December 14, 2023.
7. Toby and Eric Penrose, joint interview with author, December 14, 2023.

INDEX

Index

Index

isomorphisms, 193
Ivanenko, Dmitri, 111

Jablonna palace complex (Poland), 111
Jansky, Karl, 140
jester's cap (tile type), 205
jigsaw puzzles, 59
John Cass College, 165
Johnson, George, 261–262
Jonsson, J. Erik, 124
Journal of Mathematical Physics, 103
"Jungle" (woods), 15, 25, 106, 300

Kalbfleisch, Charlie, 90
Kendal, Marilyn, 43
Kennedy, John F., 91, 121–125
Kepler, Johannes, 61, 62, 88, 202, 272
Kerr, Roy, 103, 107 (photo), 129
Kerr solution, 103
Kestelman, Hyman, 58
Khalatnikov, Isaak Markovich, 143, 147
Kimberly Clark Ltd, Pentaplex suing, 257
King's College, Penrose, Roger, working
    at, 104
Kingswood Restaurant, 64, 83
kite tiles, 210 (fig.), 211, 219
Kochen, Simon, 209
Kodaira, Kunihiko, 69

Lake Huron, 90
Layzer, David, 156
Leathes, John Beresford (grandfather), 19
Lifshitz, Evgeny, 143
light, speed of, 121, 139
light cones, 51, 109–110
    conformal invariance exemplified by, 96
    diagram of, 52 (fig.)
    future, 52, 121, 125, 127
    "now" contained in, 121, 125
light rays, moving through space-time, 119
light vector, 51
Little Boy (bomb), 45
logic, mathematical, 76
London (England), 47, 156. *See also*
    University College, London
    American School in, 82

Austin compared with, 118
Blitz impacting, 32
Daniels, J., commuting to, 200, 204
Golders Green in, 47
Medical Research Council of, 169
Penrose, O., leaving, 64
Royal Free Hospital in, 21
Stanmore in, 103–105
London (Ontario), 21, 42–43, 45
London South Bank University, coat of
    arms for, 201 (fig.), 202
London Times (newspaper), 291
London Zoo, 80
low-entropy universe, 264–265

MacCallum, Malcolm, 183
Mach's principle, 74
Manchester University, 77
Mandelstam, Stanley, 111
many-worlds interpretation, of quantum
    mechanics, 99, 259–260
marriage, Penrose, Roger, commenting
    on, 292–293
Marshall, Frances, 21
Martin, Fiona, 284
Mason, Lionel, 251–252, 272, 277
Massachusetts Institute of
    Technology, 154
material scientists, discoveries by, 164
math, Penrose, Roger, struggling in, 36
Mathematical Games (column), 219
Mathematical Institute, at Oxford
    University, 205, 209, 234, 256,
    280, 301
mathematical logic, 76
*The Mathematical Proceedings of the
    Cambridge Philosophical Society*
    (journal), 88
mathematics, 56, 232. *See also* geometry
    limitation of, 76
    Penrose, Oliver, working with Penrose,
        Roger, on, 37–38
    Penrose, Roger, pursuing, 57
    pure, 57, 62, 66, 74–75, 77, 87
    as "sex substitute," 59–60
    World War II changing, 68

330

CREDIT: VINCENZO PIETROPAOLO

Patchen Barss is a Toronto-based science journalist who has contributed to the BBC, *Nautilus* magazine, *Scientific American*, and the Discovery Channel (Canada), as well as to many science and natural history museums. His previous books include *The Erotic Engine: How Pornography Has Powered Mass Communication, from Gutenberg to Google* and *Flow Spin Grow: Looking for Patterns in Nature*.